IDIOT'S GUIDES.
AS EASY AS IT GETS!

D0607723

# Environmental Science

by James Dauray, MEd

ALPHA

A member of Penguin Group (USA) Inc.

## ALPHA BOOKS

Published by Penguin Group (USA) Inc.

Penguin Group (USA) Inc., 375 Hudson Street, New York, New York 10014, USA • Penguin Group (Canada), 90 Eglinton Avenue East, Suite 700, Toronto, Ontario M4P 2Y3, Canada (a division of Pearson Penguin Canada Inc.) • Penguin Books Ltd., 80 Strand, London WC2R 0RL, England • Penguin Ireland, 25 St. Stephen's Green, Dublin 2, Ireland (a division of Penguin Books Ltd.) • Penguin Group (Australia), 250 Camberwell Road, Camberwell, Victoria 3124, Australia (a division of Pearson Australia Group Pty. Ltd.) • Penguin Books India Pvt. Ltd., 11 Community Centre, Panchsheel Park, New Delhi—110 017, India • Penguin Group (NZ), 67 Apollo Drive, Rosedale, North Shore, Auckland 1311, New Zealand (a division of Pearson New Zealand Ltd.) • Penguin Books (South Africa) (Pty.) Ltd., 24 Sturdee Avenue, Rosebank, Johannesburg 2196, South Africa • Penguin Books Ltd., Registered Offices: 80 Strand, London WC2R 0RL, England

International Standard Book Number: 978-1-61564-295-3
Library of Congress Catalog Card Number: 2013938007

15  14  13      8  7  6  5  4  3  2  1

Interpretation of the printing code: The rightmost number of the first series of numbers is the year of the book's printing; the rightmost number of the second series of numbers is the number of the book's printing. For example, a printing code of 13-1 shows that the first printing occurred in 2013.

*Printed in the United States of America*

**Note:** This publication contains the opinions and ideas of its author. It is intended to provide helpful and informative material on the subject matter covered. It is sold with the understanding that the author and publisher are not engaged in rendering professional services in the book. If the reader requires personal assistance or advice, a competent professional should be consulted.

The author and publisher specifically disclaim any responsibility for any liability, loss, or risk, personal or otherwise, which is incurred as a consequence, directly or indirectly, of the use and application of any of the contents of this book.

Most Alpha books are available at special quantity discounts for bulk purchases for sales promotions, premiums, fund-raising, or educational use. Special books, or book excerpts, can also be created to fit specific needs. For details, write: Special Markets, Alpha Books, 375 Hudson Street, New York, NY 10014.

**Publisher:** *Mike Sanders*
**Executive Managing Editor:** *Billy Fields*
**Executive Acquisitions Editor:** *Lori Cates Hand*
**Development Editor:** *John Etchison*
**Senior Production Editor:** *Janette Lynn*
**Illustrator:** *Brittany Breaux*

**Cover Designer:** *William Thomas*
**Book Designers:** *William Thomas, Rebecca Batchelor*
**Indexer:** *Tonya Heard*
**Layout:** *Ayanna Lacey*
**Proofreader:** *Gene Redding*

# Contents

## Appendixes

# Introduction

This book is designed to give you a primer on some of the most important concepts and issues in a very diverse, complex, and multifaceted field: environmental science.

Environmental science is a multidisciplinary study. There are a lot of concepts from many different fields you'll need to know in order to begin unraveling some of the big issues that face our society today. In this book, we'll explore lessons we've learned from past mistakes, current controversies being debated, and some of the future implications of our choices.

## How This Book Is Organized

**Part 1, The Basics of Environmental Science,** is designed to provide the most fundamental concepts the reader needs to fully appreciate and understand the ideas presented throughout the rest of the book. This part contains general science background that's relevant to environmental science, which is itself an interdisciplinary science.

**Part 2, The Biosphere,** explores the vast diversity of life and resources present on Earth. The reader will learn about the different types of species, how they evolved, why they look the way they do, and where they live. This part makes relatively little mention of human effects—that's for later, as the second half of the book explores the different ways these ecosystems and resources have been changed by human growth and needs.

**Part 3, Meeting Human Needs,** begins the gradual introduction of human influences on the ecosystems, species, and resources throughout the world. We begin by examining human population growth, projections of our growth, and how this growth varies from country to country. The focus then moves to human needs and human health, beginning with an overview of the food production system and ending with the different living and nonliving components in the environment that can be detrimental to human health.

**Part 4, A Look at Energy,** covers fossil fuels, alternative sources, and renewable, sustainable energy. Production of electricity, fuel, and heating gas is one of the single largest sources of environmental degradation discussed throughout the rest of the book. In other words, this is the "source of all evil" when it comes to many environmental issues.

**Part 5, Consequences,** looks at some of the hidden prices we pay for our level of consumption and the resources needed to support our population. These impacts are divided into air and water pollution, with the global climate change chapters directly following air pollution (since it's a direct result of carbon dioxide emissions).

**Part 6, Hope for the Future,** discusses specific changes and technologies that are vital to the human population finding a way to live on Earth sustainably. Many of the concepts from throughout the book will circle back here as we look for solutions.

## Extras

Within each chapter, there are a few different types of sidebars to highlight some important events, definitions, or even relevant quotes that highlight what we're discussing.

**DEFINITION**

These sidebars explain significant and possibly unfamiliar terms that are referenced in the section. These are important ideas to know and understand to make sense of the bigger concepts in the chapter.

**A LOOK BACK**

These sidebars discuss a specific event or time period when the concept we're covering was particularly important. Taking lessons from this history is important to avoid repeating mistakes of the past.

**CASE STUDY**

If there's a specific current event or experiment that's especially relevant to the current topic, this is highlighted in this sidebar. Applying basic concepts to new examples is a great way to sink the ideas into your long-term memory.

## Acknowledgments

To all my students, who continue to inspire me.

To my teachers, who set the standard I aspire to.

To my parents, who made me what I am.

To my wife, who gives me the support to keep working and learning.

To my son, who gives me hope for the future.

## Special Thanks to the Technical Reviewer

*The Idiot's Guides: Environmental Science* was reviewed by an expert who double-checked the accuracy of what you'll learn here, to help us ensure this book gives you everything you need to know about environmental science. Special thanks are extended to Christopher Klinger.

## Trademarks

All terms mentioned in this book that are known to be or are suspected of being trademarks or service marks have been appropriately capitalized. Alpha Books and Penguin Group (USA) Inc. cannot attest to the accuracy of this information. Use of a term in this book should not be regarded as affecting the validity of any trademark or service mark.

# The Basics of Environmental Science

Part 1 is designed to provide the most fundamental concepts you'll need to fully appreciate and understand the ideas presented throughout the rest of the book.

In the following chapters, I share a quick overview of the biggest environmental problems, along with some of the different philosophies regarding how humans live on the planet. I give you a little historical context: how human society has changed since the time of the hunters and gatherers, and how our impact on the environment has grown. I also touch on some of the regulations that have been passed to preserve our natural resources and the agencies in charge of enforcing these laws.

Finally, I provide general science background that's relevant to environmental science. Environmental science is an interdisciplinary science and requires a basic understanding of a few concepts in chemistry, physics, and the nature of science itself.

Nonrenewable resources are the most likely to be exhausted. When metals are mined from the Earth's crust, they won't regenerate, at least not within the time frame of human reference. The amount of iron, lithium, or any other metal available within the Earth is *finite*.

Renewable resources are different, as they will replenish over time. A good example of a renewable resource is water. Human society consumes tremendous amounts of water for drinking, farming, and industrial use, but it eventually finds its way back into the environment and is reused.

Don't assume, however, that renewable resources can be used indiscriminately. Renewable resources are often depleted so quickly that they're unable to replenish at the same rate.

## Air and Water Pollution

In addition to resources being consumed, they can also be degraded by human activities. This is *pollution*. This degradation can impact the most basic elements of living ecosystems: soil, air, and water. Pollution is considered a fundamental environmental issue because these resources are so basic and vital to all life forms on Earth.

Pollution is primarily released into air or water. This spreads the problem, as each of these substances moves through the Earth readily and quickly.

After it's released, air pollution tends to move in the direction of the prevailing wind currents. The New England states are sometimes referred to as "America's tailpipe" because much of the pollution produced within the Rust Belt states of the Midwest finds its way there. Industrial pollutants have been found in rural areas miles away from any power plants or factories.

Water pollution can also move, but not as freely as air. If pollutants are released into a lake or pond, they'll largely stay there. However, if they're released into a river, they can affect communities and ecosystems downstream, eventually even finding their way into the ocean.

## Loss of Biodiversity

Visit a public place and look around you. Consider the range of ages, ethnicities, and religious beliefs among the people you see there. These differences describe the diversity of the group. Now look at a natural ecosystem. Consider the different types of trees, shrubs, fungi, insects, mammals, birds, and other organisms you see or hear. This is the *biodiversity* of that ecosystem.

The number of species that exist on Earth is unknown, but it is estimated to be in the tens of millions. With so many species living in the same space, competing for the same resources, extinction of some species is inevitable.

**CASE STUDY**

A famous paper published by Sepkoski and Raup in 1982 identified the "big five" major extinction events believed to have occurred during the history of life on Earth. The most recent is called the Cretaceous-Tertiary extinction event, as it separates those two geologic periods. Causes of this event are hypothesized to be huge amounts of dust entering the atmosphere from one or more asteroid impacts and increased volcanic eruptions. The dust blocked out enough sunlight to interfere with photosynthetic organisms. As much as 75 percent of all species may have gone extinct during this event.

Extinction occurs when a species completely dies out. This can happen naturally as other species evolve better ways to compete for the same resources or from a sudden catastrophic change, such as a volcanic eruption or meteor strike. Sudden, large extinction events are pretty rare, however. Assuming no great catastrophes, a typical ecosystem will experience an extinction rate of one mammal every 200 years.

How do humans influence this rate? Consider the example of Australia and the nearby smaller island of Tasmania.

Australia has seen the extinction of 54 species, including mammals, birds, and amphibians, since the late eighteenth century. What catastrophic event underlies these extinctions? The first Europeans settled Australia in 1788, opening the way to increasing human population, consumption of natural resources, and the introduction of non-native species into the continent.

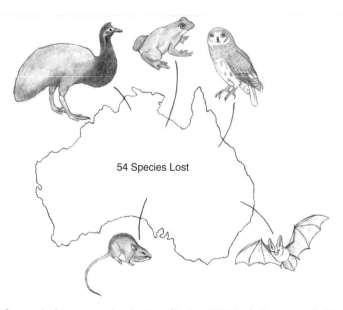

54 Species Lost

*Fifty-four species have gone extinct in Australia since 1788, including mammals, birds, and amphibians. This is primarily due to the influence of European settlers.*

One of the single most damaging decisions made by the first European settlers of Australia was the introduction of the European rabbit. The ecosystem was so ideal for the rabbits that their population exploded, stripping the land of vegetation the native species needed. The native animals, unable to compete with the rabbits, gradually disappeared. Suddenly, extinction rates normally associated with extremely rare catastrophic events occurred due to seemingly innocent human decisions.

The island of Tasmania, on the other hand, has not seen this level of destruction. The island was largely spared the huge infestations of invasive species, and it has a much lower extinction rate as a result.

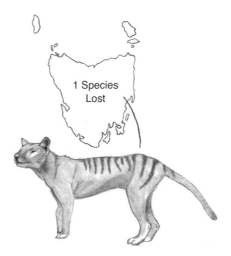

*Tasmania has seen only one extinction, the Tasmanian tiger, in the last two centuries. This matches the background extinction rate of mammals that would be expected outside of human influence.*

The other important aspect about extinction is its permanence: once a species goes extinct, it will never return. Other species may take its place within an ecosystem. Species such as the Desert bandicoot in Australia will never return, although other closely related mammals may remain.

# Ethics and Decision Making

Debates have sprung from the number of pressing environmental issues that have resulted from human population growth. With any decision that uses a natural resource, releases a pollutant, or endangers an ecosystem, the human needs are weighed against the effects on nature.

Given the complex nature of the environment and the long list of human needs, there will be multiple sides in any environmental debate or controversy. The position you take in regards to one of these issues is a reflection of your environmental ethics.

One of the earliest and most well-known environmental conflicts was the proposed construction of a dam in the Hetch Hetchy Valley in Yosemite National Park. The city of San Francisco was seeking a more reliable source of water to support a burgeoning population in the early 1900s, and the topography of Hetch Hetchy was ideal for creating a reservoir. Should the dam be constructed, despite the seeming violation of the intended purpose of the national park?

## Anthropocentrism

The word *anthropocentrism* literally means "human-centered." Individuals with this philosophy believe the Earth and its resources exist primarily for human needs. According to this line of thinking, humans dominate the Earth and should develop to benefit the population as much as possible.

*The anthropocentrism environmental worldview places human needs above those of all other organisms and the environment as a whole.*

For most of human history, this has been the most prevalent philosophy. This isn't necessarily a bad or wrong viewpoint to take, because the human population does have a set of basic needs that must be met. Is it possible to be an anthropocentrist and still environmentally responsible?

A good example of this is Gifford Pinchot, the first chief of the U.S. Forest Service. He famously wrote, "Where conflicting interests must be reconciled, the question shall always be answered from the standpoint of the greatest good of the greatest number in the long run." In other words, human needs are still a priority, but the needs of future generations are to be considered as well as the immediate ones.

Presented with the Hetch Hetchy debate, what would anthropocentrists do? Certainly, they would vote to construct the dam. The benefits to human society are numerous, from the improved water supply to San Francisco and surrounding areas to the tourist-friendly artificial lake that would fill in behind the dam.

## Biocentrism and Ecocentrism

The center of the *biocentrism* philosophy is all life, rather than just humans. We are seen as but one species among millions. Essentially, every organism is believed to have its own value or purpose to the greater ecosystem or to Earth itself.

The ecocentrism philosophy is centered on nature as a whole. Value is not placed simply on living species, but on all aspects of the environment. This would include nonliving characteristics such as climate, air quality, and water quality.

 **DEFINITION**

**Anthropocentrists** make decisions strictly based on human needs, both short-term and long-term. **Biocentrists** look at the needs of other species besides humans, while **ecocentrists** consider the importance of maintaining entire ecosystems in the most pristine condition possible.

*The ecocentrism environmental worldview considers humans as part of the environment. The well-being of humans as well as other species enters into any decision.*

Making a distinction between biocentrism and ecocentrism is difficult, as individuals from each philosophy often come to similar conclusions. John Muir, founder of the Sierra Club, argued that the purpose of the national park was to preserve the ecosystems in their natural form. To change or alter this ecosystem, in his view, was sacrilegious.

The differences among anthropocentrists, biocentrists, and ecocentrists came into full focus during the Hetch Hetchy debate. Gifford Pinchot argued for the water needs of San Francisco's citizens. John Muir pled with Congress to leave Yosemite in its untouched, natural state.

As for the Hetch Hetchy decision, human needs won out. In 1913, the dam was approved for construction. The reservoir remains in use today and accounts for about a quarter of San Francisco's municipal water supply.

# Same Planet, Different Worlds

To further complicate the list of environmental issues, debates, and philosophies, the types of issues faced by countries around the world are not all the same. Each country has a different level of wealth, politics, and available resources.

One of the biggest factors influencing the environmental problems faced by any given country is its economic status. Wealthy countries and poor countries have huge differences in the sizes and composition of their respective populations.

## Developing Countries

Developing countries are those with lower gross national products, lower literacy rates, shorter average life spans, and more rapid population growth. About 80 percent of the world's population falls within this definition, although they consume resources at a much lower rate. Examples include much of Central and South America and Africa south of the Sahara Desert.

Environmental policy in developing countries is very much anthropogenic and short-term. This is often out of necessity, as the population may be on the brink of not surviving. The most common causes of death in these countries are disease and malnutrition.

Resource depletion is very common in developing countries. As an example, both forests and native animal species are considered renewable resources. Each will regrow and repopulate over time. However, forests are often cut down at such staggering rates that regrowth cannot keep up with this destruction. Overhunting is also a common problem, especially if specific parts of the animal are highly valued. Rhinoceros horns, for example, are ground and used as herbal remedies in many Asian countries.

**CASE STUDY**

A revolt following European colonization of Haiti led to the island being primarily populated by former African slaves. In the nineteenth century, any land that was fertile was allocated to the former slaves, who are now farmers. As the farmers passed their land to their children, it was divided into smaller and smaller parcels. The subsequent generations began to farm much more intensively and overused the land. Over time, the soil could no longer support crops. The farmers then moved to surrounding hillsides. Trees were cut down and sold, and the soil was used until it again became infertile. Now, much of the country cannot support crops, deforestation is prevalent, and famine is commonplace.

## Developed Countries

In most ways developed countries are mirror opposites of developing countries. They have higher gross national products, higher literacy rates, longer average life spans, and much slower population growth. Only about 20 percent of the world's population lives in these countries, but they consume resources at a much greater rate.

Developed countries—those with the highest gross domestic products per capita in the entire world—include Singapore, Norway, the United States, and the United Arab Emirates.

To get a glimpse into some of the demographic differences between the developed and developed world, consider this comparison table of statistics taken as of 2011. The two countries chosen, the United States and Bangladesh, are very much at opposite ends of the economic spectrum.

### Comparison of Developed and Developing Countries

| Measurement | United States | Bangladesh |
| --- | --- | --- |
| Life expectancy in years | 78.2 | 48.3 |
| Population growth rate | 1.0% | 2.7% |
| Gross national product per capita | $48,890 | $910 |
| Energy use per year per capita | 10,381kWh | 35kWh |
| Literacy rate | 99% | 38% |

# The Struggle for Sustainability

The ultimate goal of environmentalism is an idea called sustainability. This isn't a new idea. In fact, in 1972, the United Nations itself created an organization whose purpose was to promote sustainable living practices. Deterioration of the environment was a problem in countries all over the world. A study group established from this effort, called the Brundtland Commission, wrote a report that contains the best and most-cited definition of sustainability.

From that paper, "Our Common Future":

> Sustainable development is development that meets the needs of the present without compromising the ability of future generations to meet their own needs.

Simply put, the idea of sustainability is for the human population to live in such a way that its own needs can be met *indefinitely*. This often is not the case, however, both in developed and developing countries.

## The "Tragedy of the Commons"

In 1968, ecologist Garrett Hardin wrote an essay titled "The Tragedy of the Commons," suggesting that the root of environmental problems was a conflict between the short-term interests of individuals and the long-term interests of the society as a whole.

A *commons* is an area that isn't privately owned; rather, it's shared by the surrounding people. A modern example of this is international waters in the oceans. These areas of the oceans do not fall within the control of any one country, and thus no laws or regulations protect them. In theory, any individual could utilize these waters.

When it comes to the treatment of a commons, individuals will often deplete or overuse it as they seek the most short-term benefit possible. Do renters behave differently from homeowners? Do public bathrooms look different from ones in private homes? Would you drive a rental car more carefully than your own car?

In the case of the world's oceans, much of the oceans suffers from pollution and overharvesting of stocks of fish. Ocean waters that fall within an individual country's control are less likely to suffer this problem, as the country is more apt to regulate fish harvests to preserve the industry in the long term.

## Ecological Footprint

Both developed and developing countries struggle to meet the definition of sustainability, but it's the wealthier countries that consume far more, potentially depriving future generations of resources. One way to measure this difference is with an ecological footprint, a measurement of the total amount of land needed to support a given lifestyle. This includes farmland for food, forests for timber, mines for minerals and fossil fuels, and so on.

Developing
Countries

*The ecological footprint in a developing country is relatively small. A greater reliance on manual or animal labor, in addition to a lack of electricity, contributes to this size.*

The ecological footprint for an average person in the United States is about 30 acres, while for an average person in India, it's closer to 3 acres.

What lifestyle differences account for this tremendous disparity? How much electricity does an average individual in each country consume? Does an average person in India have access to the same array of personal electronics—laptops, cell phones, and video game systems? What are the differences in diet? Many in India have a primarily vegetarian diet, which has much lower resource costs to produce. What percentage of India's population owns a car and drives frequently? Each of these adds up to a huge discrepancy in resource consumption.

This contrast in lifestyles clearly illustrates the differences in consumption between the two types of countries and the tremendous number of changes needed to create a more sustainable society.

The number of environmental problems facing us is vast, but not impossible to solve. Many of them stem from the same issue—that is, a tendency we as humans have to think only in the short term. If we could begin to consider the well-being of other species and societies, and shape our lifestyle accordingly, our species would be living much more sustainably.

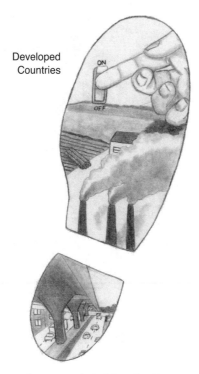

Developed
Countries

*Developed countries have a much greater ecological footprint. Easy access to electricity, gasoline, and machines and an overall higher standard of living contribute to this size.*

## The Least You Need to Know

- Resource depletion, loss of biodiversity, and pollution are considered the top environmental problems.
- Problems in developing countries stem from poverty—deforestation, famine, malnutrition, and poaching—while problems in developed countries stem from overconsumption of resources.
- The "Tragedy of the Commons" essay explains the cause of environmental problems as a conflict between short-term interests of individuals and long-term interests of entire populations.
- Sustainability is an ideal way of living so that a population can continue indefinitely.

# Environmental History

The types of interactions people have had with the environment have changed tremendously since the appearance of the earliest humans. As we've developed new lifestyles, grown our population, and introduced new technologies into the world, our impact on the environment has grown exponentially. Trying to keep pace with this impact by writing laws and developing cleaner technology has been a tremendous challenge for the human society.

In this chapter, we take a quick tour through some of the most important turning points in environmental history. As humans, we have a tendency to learn from hindsight—from the mistakes we've made in managing the world's resources. I share examples of some of the biggest environmental disasters throughout this book.

We humans have learned from the mistakes we've made in working with the environment. We have entire government agencies tasked with monitoring pollution, resource use, and the impacts of human decisions on ecosystem health. Laws are in place to protect the air we breathe, the water we drink, and other organisms. Many of these laws are relatively recent, however, and a great deal of history occurred without the modern understandings we take for granted.

## In This Chapter

- The evolving interactions between humans and the environment
- Changes that resulted from the most significant environmental movements
- Environmental events that changed how we view the world
- Agencies that monitor and protect our land, air, and water

# The Growing Impact of Humans

Humans, much like any other species on Earth, live by using different resources taken from the environment. This ranges from harvesting plants to hunting animals to burning wood or coal for fuel. What separates the human species from other animals is our intellect—our ability to adapt, to use tools, to extract resources, and to survive at a much greater rate than might normally be expected.

This hasn't always been the case. Our ability to communicate, to design new technology, and to adapt to remote and harsh ecosystems has developed gradually over the thousands of years humans have inhabited Earth. As *Homo sapiens,* the human species, has evolved, so has the magnitude and nature of our impact on the Earth.

## A Brief History of the Earth

The Earth is believed to be about 4.5 billion years old. Compare this with the oldest known *Homo sapiens* fossil, which is only about 200,000 years old. A great deal of time passed on the Earth before even the ancestors of our species first appeared.

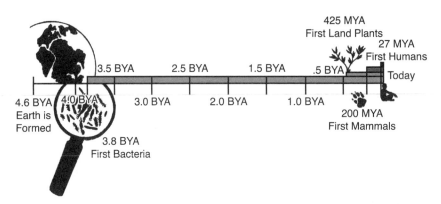

*Human history occupies a relatively small portion of the history of the Earth. Even the earliest mammal fossil is only 220 million years old—$\frac{1}{20}$ of the Earth's total history!*

The first single-celled organisms are believed to have formed around 3.5 billion years ago in the oceans. In fact, all life is believed to have been limited to the oceans for the next 4 billion years. Plants as we know them finally appeared on land, followed by insects, reptiles, spiders, and amphibians. The rise and fall of the dinosaurs was followed by the emergence of many new mammal species, including the first humanoids.

During these billions of years, the Earth went through many natural climate cycles of warming and cooling through a series of *ice ages*. Glaciers formed during *glacial periods* and receded during *interglacial periods*. Species evolved and went extinct. Around the time of the earliest known *Homo sapiens*, the Earth was entering the last glacial period.

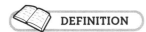 **DEFINITION**

> **Ice ages** are periods of time lasting millions of years. Within an ice age, **glacial periods** are shorter intervals of time that have overall colder temperatures (and more glaciers) across the Earth. **Interglacial periods** are the opposite: intervals of time within an ice age when the Earth is warmer overall and the glaciers recede. We are currently in an interglacial period within the most recent ice age.

## Hunters and Gatherers

The earliest humans were nomadic. Their lifestyle was much different from what we associate with modern society. They didn't grow their own food, domesticated animals largely didn't yet exist, and settlements weren't built to be permanent. Like other animals that hunt in packs or graze in herds, humans lived in groups that were constantly on the move in search of resources.

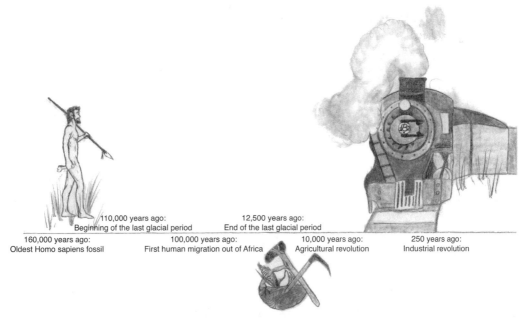

110,000 years ago:
Beginning of the last glacial period

12,500 years ago:
End of the last glacial period

160,000 years ago:
Oldest Homo sapiens fossil

100,000 years ago:
First human migration out of Africa

10,000 years ago:
Agricultural revolution

250 years ago:
Industrial revolution

*The way of life for humans has changed tremendously since the hunter-gatherer period, with much greater resource consumption since the Industrial Revolution.*

Fossil records give evidence to the theory that humans originated in Africa. Much of this continent is dry and desertlike all year or very seasonal in regard to precipitation. As a result, most animals that reside within African ecosystems are forced to periodically travel great distances in search of food as the seasons shift.

During the wet season, when precipitation levels are highest, plant life is abundant. The earliest humans were likely able to take advantage of this by gathering edible plant material. The abundance of plant material would attract animals as well. Early human tribes hunted these animals as a source of meat.

The plains of Africa change dramatically during the dry season. The rain becomes incredibly scarce, much more similar to a desert ecosystem. Many of the grasses and wildflowers dry up and become dormant. The herds of plant-eating animals, easily supported during the wet season, suddenly face a real danger of starvation.

In response to the change in season, many of the animals migrate to other areas of the continent searching for enough food to survive. With the absence of plant material to gather and animal herds to hunt, human tribes would move as well. Villages, towns, and cities as we now know them simply were not possible.

 **A LOOK BACK**

Why did humans leave Africa and begin spreading throughout Asia, Africa, and eventually the Americas? The most likely reason is a pretty mundane one: they were looking for new sources of food. Conditions in Africa may have become difficult due to drought or increased competition for available food sources. This, incidentally, is the same reason why many animal species have spread throughout continents. As the world was entering the last glacial period, a great deal of water was trapped in the polar ice caps, resulting in overall lower sea levels. This may have allowed early humans to cross bodies of water that are much wider today, such as the Red Sea.

## The Agricultural Revolution

For much of early human history, the world was in the middle of the last glacial period. Much of northern Europe, Asia, North America, and southern South America either were covered in ice or had permanently frozen soil. Sea levels were much lower, and the overall climate was dry. Any kind of organized agriculture would have been near impossible in most of the world because the rainfall wasn't consistent or predictable.

The last glacial maximum, the point at which the ice sheets were at their peak, occurred between 20,000 and 25,000 years ago. As the glaciers began to recede, human populations began domesticating plant species. While there are several competing theories about what triggered this shift away from a hunter-gatherer lifestyle, the changing climate likely played a big role. Human settlements became permanent, resembling what we now call villages and towns. Animals were also domesticated, freeing humans from having to follow the migrating herds.

The Agricultural Revolution marked the beginning of humans independently influencing their surrounding environment. New breeds of plants and animals were developed, emphasizing traits like size and milk production that benefited the early farmers. Ecosystems such as grasslands were converted into farmland. The establishment of permanent settlements resulted in an increased human population, higher population densities, and the first forms of governance. These influences continued to grow at a steady rate through prehistory and into the Middle Ages.

## Ancient Civilizations

Human society evolved a great deal during these passing periods. Empires rose and fell, and the species spread out to every continent except Antarctica. Their impacts on the environment increased as their population grew.

During the hunter-gatherer period, humans had little impact on the environment any different from the animals. Moving forward from the Agricultural Revolution, this began to change. Large cities and dense populations began to appear throughout Europe and Asia.

The earliest environmental problems also began to appear during this time. Ancient Romans primarily relied on burning wood for heat, cooking, blacksmithing, and so on. With all the burning of wood, air pollution became prevalent. Additionally, larger and larger amounts of forests around Rome were clear-cut. Water pollution was also commonplace, as latrines were flushed out through channels into nearby rivers.

 **A LOOK BACK**

In 61 C.E., the philosopher Seneca wrote about the air conditions in Rome, "As soon as I had gotten out of the heavy air of Rome and from the stink of smoky chimneys thereof, which, being stirred, poured forth whatever pestilential vapors and soot they had enclosed in them, I felt an alteration of my disposition." He felt physical relief in escaping the pollution of the city!

## The Industrial Revolution

Wood was the central source of fuel in ancient and medieval civilizations. Wood had to be burned to craft metal, cook, and perform all the other artisanal tasks within the city. Fuel wood was also needed as a source of heat during the colder months. One of the earliest civilizations to use coal was in Britain during the Middle Ages (the eleventh century). However, its use was limited to smiths in regions where coal was easily accessible.

The eighteenth century marked the beginning of machine-based manufacturing. Coal began to replace timber as a primary source of energy. Inventions such as firearms, steam power, and electricity had profound impacts on human society—and on the environment.

While humans had always produced pollution in small amounts, it was primarily through the burning of timber. The Industrial Revolution, however, brought about the burning of more fossil fuels, especially coal. Oil was utilized to produce synthetic products such as plastics. Fertilizers, pesticides, herbicides and other new chemicals were introduced into the environment. Picture the pollution issues the ancient Romans caused mixed with smoke from burning coal for machines.

Compounding these new impacts was the tremendous growth of human population. It took thousands of years to reach the first billion people, but the second billion was achieved in only a century. Ecosystem loss, degradation in water and air quality, and extinction rates all began to climb in earnest during this period, and they continued to do so nearly unfettered for decades. This, after all, was seen as necessary for human progress.

# Progress, One Step at a Time

The consequences of the rapid growth of the Industrial Revolution began to be apparent in the United States in the late nineteenth century. The first public environmental concern was actually the loss of natural wilderness.

By this time, the frontier of the West had largely been settled. All censuses in the United States taken prior to 1890 contained a frontier line—a border drawn on maps beyond which the population density was extremely small. The presence of this frontier created the drive for expansion, exploration, and the creation of new settlements.

This all changed with the 1890 census. The director of the Census Bureau declared the frontier closed. There would be no frontier line on the maps from this point forward, as the population of the United States had grown and spread so much, so quickly. All of a sudden, the notion of limited westward expansion was shattered.

## The Progressive Era

As the concept of the U.S. frontier began to fade in the late nineteenth century, much of the public began to notice that the unique, untouched ecosystems were disappearing as well. A series of land preservation laws was passed; environmental organizations such as the Sierra Club were founded; and Theodore Roosevelt, one of the most famous conservationists, became president.

While the laws and regulations of the Progressive Era centered around land conservation, the modern environmental movement focused more on issues with pollution. What changed during this era? Several national parks were created, including Yosemite, Crater Lake, and Sequoia. These parks were intended to protect the ecosystems from development, logging, and hunting. This is why the proposed Hetch Hetchy Dam in the early twentieth century was so controversial. The land had been set aside specifically so it wouldn't be altered in any way! As of 1916, these parks are managed by the *National Parks Service.*

The Forest Reserve Act of 1891 created millions of acres of publicly owned forest lands. This prevented the forests from being claimed for farmland by homesteaders. Since 1905, the agency in charge of administering this land has been the *United States Forest Service (USFS).*

Clearly, the priority for the environmentally minded of this era was land preservation and conservation. Issues like water and air pollution largely weren't addressed, with the exception of the Rivers and Harbors Act of 1899, which banned waste disposal in navigable waters. The purpose of this law was not to protect wildlife or human health; rather, it was to ensure the waters were unobstructed for boats passing through.

## A LOOK BACK

The Dust Bowl of the 1930s was the worst drought and biggest agricultural disaster that ever occurred in the United States. Little care was given to the soil of the temperate grasslands of the Midwest as the deep-rooting native plants were replaced by much shallower-rooting corn, soybean, and wheat crops. This was the beginning of a lesson in land conservation that had to be learned the hard way.

The soil began to be depleted of nutrients due to intensive agriculture. Then a multiyear drought pulled so much moisture out of the soil that the wind began blowing it away in massive dust storms. The native root systems that normally would have held the soil in place were gone.

To top everything off, an infestation of grasshoppers, desperately looking for any edible plant material, finished off any crops that had managed to survive the drought.

## The Birth of the Environmental Movement

Great strides in land conservation were made during the Progressive Era, but several major pollution events during the mid-twentieth century began to create a sense that another great environmental crisis was impending. The price of the progress of the Industrial Revolution was becoming more and more apparent.

The most visible of these problems was air pollution. Donora, Pennsylvania, south of Pittsburgh, was covered in thick smog from its zinc factories that sickened thousands. London had a similar, even more deadly air pollution incident that stemmed from the use of coal to heat homes. We'll cover these events in more detail in Chapter 19.

Questions were beginning to be raised about the excessive use of synthetic pesticides. The pesticide DDT had been used during World War II to kill mosquitoes and control the diseases they spread. A Nobel Prize was even awarded to the chemist who discovered that DDT was effective against insects. However, Rachel Carson's book *Silent Spring* documented the toxic buildup of the DDT insecticide that had been occurring unseen in all levels of the food chain. DDT was threatening the populations of large predatory birds, including the bald eagle.

Throughout the Industrial Revolution and much of the twentieth century, factories and power plants commonly (and legally) dumped their solid and liquid wastes into nearby soil, rivers, and lakes. This pollution had become so pervasive in Ohio that in 1969, the Cuyahoga River actually burst into flames due to a massive buildup of oil and other chemicals in the water.

## Modern Environmentalism

Public outcry grew in earnest in response to these terrible environmental disasters. As a result, the U.S. government created a series of environmental legislative acts.

The Clean Air Act, passed in 1970, placed restrictions on the most prolific and damaging forms of air pollution. The *Environmental Protection Agency* (*EPA*) was created the same year, to provide a means of enforcement. The Clean Water Act of 1972 enabled the EPA to regulate the discharge of pollutants into water sources. The Safe Drinking Water Act was enacted in 1974 to create a system for testing and monitoring water from rivers, lakes, reservoirs, and springs. The focus of these two laws was much different from what had passed during the Progressive Era. Now, it was on the importance of clean water to the health of humans and other organisms.

These laws were only the beginning and have since been modified with additional requirements, new technologies, and improved monitoring. (I discuss the present state of each law further in Chapters 19 and 21, on water and air pollution.)

 **DEFINITION**

> The **Environmental Protection Agency** (**EPA**) enforces regulations based on laws passed dealing with air, water, soil, and species preservation. The **United States Forest Service** (**USFS**) administers publicly owned forests and grasslands. The **National Parks Service** (**NPS**) manages all national parks and federally designated monuments and historical landmarks.

Additions to these laws have created the framework for most of the environmental legislation in place today. We've made a lot of progress since the dirty, acrid days of the Middle Ages. Beginning with land preservation and moving through protection of air and water quality, our society is taking a lot of key steps toward preserving our resources for the future.

## The Least You Need to Know

- The nomadic hunter-gatherers made environmental impacts similar to any animal species—hunting other animals, migrating, and consuming plants.
- The Agricultural Revolution spurred by the end of the last glacial period marked the beginning of permanent settlements and animal and plant domestication.
- The Industrial Revolution brought a massive increase in the use of fossil fuels—mostly coal—and with it, a great deal of pollution.
- The primary focus of legislation during the Progressive Era of the late nineteenth and early twentieth centuries was land conservation and preservation.
- The modern environmental movement ushered in a focus on dealing with air and water pollution.

# Scientific Principles

Environmental science, biology, physics, chemistry—while each of these disciplines focuses on different aspects of nature and the universe, they all follow the same basic rules. They're all based on verifiable evidence that has been collected through standard procedures of observation and experimentation. In other words, they are all *science*.

Science can be a frustrating study at times, because there are no 100 percent proven explanations. Instead, science is a collection of theories and hypotheses supported by a great deal of evidence. Each of these explanations is fluid and dynamic. Each can be changed and altered, in ways both big and small, when new evidence is collected.

All of the ideas in this book are based on current research and experimentation. This chapter gives you an overview of how this process works, how we know the data is reliable, and a few fundamental concepts from other scientific disciplines you'll need going forward.

## In This Chapter

- False science and how it's prevented
- The importance of the scientific method
- A look at the structure of an atom
- The essential laws of energy and matter

# Bias, Deceit, and False Science

A great deal of our scientific understanding has been developed relatively recently, within the last 200 years. Prior to that, many accepted understandings of the world were based on incomplete observations and unverified claims.

For example, consider the notion that the world is flat. This seems like a perfectly rational concept. Walk outside and look around—everything is flat, right? Of course, we now know the world is round. It only *seems* flat from the limited perspective of an individual observer. This is an example of *pseudoscience,* or false science. The claim was presented at the time as science but wasn't actually based on research as we now define it.

Often, science relies on *quantitative data*—actual numerical measurements taken during an experiment. Pseudoscience doesn't use quantitative data, but rather *qualitative data;* it relies on assumptions and untested explanations.

 **DEFINITION**

**Pseudoscience** is a claim that appears or presents itself to be objective science, but is not. **Quantitative data** is numerical measurements taken as part of a scientific experiment. Compare this to **qualitative data,** or non-numerical observations such as color change or how strong something tastes.

One of the earliest documented quantitative scientific experiments was performed by Francesco Redi, a seventeenth-century Italian physician. During his time, a theory known as spontaneous generation was used to explain many phenomena involving the sudden appearance of life forms where none had existed before. Put simply, the theory stated that life could be created from non-living matter. For example, one "recipe" for generating life involved placing wheat and soiled linens in a dark room. A few days later, rats would appear!

Redi had a quality found in all good scientists: he was skeptical. He felt that if this theory were legitimate, it should be observable in a controlled setting. He decided to test one of the recipes by taking two pieces of old meat and placing one in a closed flask and the other in an open flask. Days later, he observed maggots growing in the open flask but nothing in the closed flask. However, it was argued that the meat needed access to air in order for life to arise!

So Redi repeated the experiment, this time placing gauze over one of the flasks. As before, maggots appeared only in the open flask. They had not been created from nonliving matter; they were hatched from eggs laid by an adult fly!

# Peer Review

Again, when scientists are presented with a new theory or claim, they're skeptical. How can you judge the difference between true science and pseudoscience? An important aspect of all scientific research is that it must be *verifiable* and *repeatable*.

Verifiable means that the data is accurate. A fellow scientist reading research findings in a published journal article may want to evaluate the data himself. He'll examine the procedures used to collect the data and look for flaws or inconsistencies. Still unconvinced, he may also attempt to perform the study himself. Ideally, the data the second scientist collects should be very similar to that from the original study. In other words, the experiment is repeatable. Looking at complementary results from two independent researchers, even the most ardent skeptic will begin to give this claim legitimacy.

 **A LOOK BACK**

Although science has changed a great deal since the flat Earth theory, fraud still exists. One of the most infamous recent examples is a study published in a well-respected British journal linking autism and childhood vaccines. An investigation concluded that the study's author altered the medical histories of all 12 of the patients cited in the study. This fraud was not uncovered for more than a decade, and was directly responsible for an increasing number of parents choosing not to have their children vaccinated.

# Bias

While some rejected studies and claims are the result of honest error or bad experimental design, they can also be the result of the researchers purposely manipulating the data. What could be the motivation for this kind of fraud? Fame and financial gain are the most likely incentives. Consider a team of doctors and researchers working on a newly developed drug. If this drug is shown to be successful, promotions, bonuses, and a lot of news coverage could all be forthcoming. Conversely, failure could derail all these ambitions.

*Bias* is most likely to be a problem when scientific data is qualitative instead of quantitative. Qualitative data is not numerical; for example, it might be an observed color change during a chemical reaction. This kind of data, by its nature, is subject to some interpretation by the scientist.

 **DEFINITION**

**Bias** is the manipulation of data from an experiment on the part of a scientist or a patient. Patient bias is often referred to as the "placebo effect," where relief from symptoms is felt simply because of the expectation of help from the drug being tested.

A psychic at a well-known research institute had documented claims of being able to manipulate the blood pressure and brain waves of patients through his own mental powers. An entire team of doctors claimed to have witnessed and documented this ability. One of the problems, however, was that the patient, doctors, and psychic were all together in the room when the tests were conducted. Bias on the part of the doctors (looking to preserve their jobs) was likely, especially given the qualitative nature of the measurements. Bias on the part of the patient was also possible; after all, wouldn't your blood pressure go up if a psychic was directing his "powers" toward you?

As part of a PBS *NOVA* episode titled "Secrets of the Psychics," an investigator was sent to Russia to test these claims. How could an experiment be designed to minimize the possible sources of bias and generate reliable data? A more quantitative method would be required.

# The Process of Scientific Discovery

Bias, fraud, poor design, and incomplete data are all roadblocks to scientific progress. Combined, these can slow down research to a crawl. One important historical example of this problem is the discovery of the atom. Today, the concept that all matter is made of atoms is one of the most fundamental in all of science, especially chemistry. This idea is actually first attributed to the ancient Greek philosopher Democritus, who lived around 400 B.C.E. However, the idea never really gained traction until the eighteenth century! What accounts for this delay?

Around the same time as Democritus, a competing idea that all matter was made of the four elements (wind, fire, earth, and water) was proposed by another famous philosopher, Aristotle. The elements theory was much more concrete and intuitive than the atomic theory. Remember, there were no microscopes, no electricity, and no periodic table in ancient Greece. Thus, Aristotle's ideas were largely accepted and formed the basis of the study of alchemy for the next 2,000 years! Throughout the Middle Ages, nearly every monarch had an alchemist within his castle busily attempting to discover how to manipulate the four elements of matter so that one form of matter, such as lead, could be transmuted into another, such as gold.

## The Scientific Method

The scientific method is a series of steps that make up all good experiments. This is one of the first concepts taught in a typical science course. The different parts include the following:

**Observation:** A condition or phenomenon that is witnessed but not fully understood. This leads to the question that forms the whole basis of the experiment or study.

**Hypothesis:** A prediction of the results of the experiment based on prior understanding or study of the subject.

**Experiment:** A set of procedures designed to directly test the predicted results from the hypothesis. Any data will be collected during this step.

**Conclusion:** The data is summarized and used to either refute or support the original hypothesis.

**Publication:** The entire experiment—including the original hypothesis, the complete set of procedures followed, the data collected, and any conclusions—is published in a journal. Other scientists in the field examine and critique the results.

Let's illustrate each of these stages with a simple example. Consider a large pond in a suburban area that's covered by a thick layer of green algae throughout the summer months. The *observed* algae overgrowth in the pond leads to the *question,* "Why is this pond so prone to algae overgrowth?" An environmental scientist, having read about similar algae blooms in the ocean, immediately *hypothesizes* fertilizer runoff is entering the water from surrounding homes. An *experiment* is conducted in which water samples are pulled from this pond as well as others in the area not suffering from the algae bloom. The levels of nitrates and phosphates in these water samples are recorded into data tables for comparison.

| Location | Nitrate (mg/L) | Phosphate (mg/L) |
|:---:|:---:|:---:|
| 1 | 0.10 | 0.02 |
| 2 | 0.10 | 0.04 |
| 3 | 0.60 | 0.10 |

*This data table displays the nitrate and phosphate measurements taken from three different locations. Sample 3, from the affected pond, has the highest levels.*

The results show elevated levels of both fertilizer components in the affected pond water, leading the scientist to conclude that fertilizer pollution is the most reasonable explanation. These results are published in the local homeowners association newsletter.

## Parts of an Experiment

All experiments are different in some ways. For example, some are conducted in nature, while others take place in controlled laboratory settings. Every experiment, however, has the same basic purpose: to test the effect of one specific variable on an outcome.

The variable that is tested or manipulated in an experiment is known as the *independent variable*. The variable that changes and is measured as data is the *dependent variable*. While these names can be easy to mix up, just determine within each experiment what the scientist is actively manipulating (independent) and then measuring (dependent).

Consider the algae bloom study discussed previously. The scientist was actively selecting water samples from different ponds. This would be the independent variable. Each of these samples contained different levels of nitrate and phosphate pollution. These measurements would be the dependent variables.

Look at the chart of the data collected from this fictional experiment:

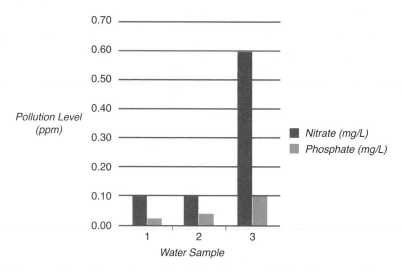

*The bar graph shows the same data as the table, but in a more visual and easy-to-analyze way. Location 3 (the polluted pond) clearly shows elevated levels of both nitrates and phosphates.*

Notice the independent variable is on the x-axis, while the dependent variable is on the y-axis. This convention is typically followed when graphing data collected from a scientific experiment.

## Combating Bias

Not all data is clear-cut, and some interpretation might be required on the part of the researchers. Consider a trial study for a drug developed to treat a disease. What would be considered successful results? Some doctors would define this as a lessening of symptoms, while others set a higher bar, looking for the near elimination of symptoms. This interpretation can be easily swayed by the personal feelings and ambitions of the researchers. This is bias. James Randi, a famous investigator of psychic phenomena, summarizes the problem of bias by saying, "Scientists have the uncanny ability to find what they're looking for."

Bias is counter to scientific progress and must be minimized. Experiments can be designed with just this in mind. Let's continue with the example of drug trials and look at bias on the part of both the patients and the researchers.

Most sick people want very badly to recover. When they're offered a treatment, they may notice a lessening of symptoms, even though the actual physical disease is still present. This is a psychological phenomenon known as the *placebo effect*. How do you separate individuals who have been successfully treated from those who simply think they have? You *blind* them. Not literally, as in blindfolding; you give one group a fake pill (placebo) that looks just like the real one, while the other group gets the real drug. If both groups report similar results, researchers know the drug was no more effective than a placebo.

Designing blind studies like this accounts for patient bias, but what about the researchers? Pharmaceutical companies will blind them as well. The researchers will be dispensed the drug, but they, like the patients, won't know whether they've been given the placebo or not. In order for this to work, there must be a few people who design the experiment and dispense the drugs. Typically, these will be high-ranking, well-paid individuals within the company whose careers are not impacted by the success or failure of the drug. This is a *double-blind* experiment. This is the most effective experimental design that minimizes bias.

 **DEFINITION**

A **blind** experiment attempts to remove bias on the part of the patients. This is often done through use of a **placebo,** a substance or procedure that closely resembles the actual treatment, but does not actually contain the active ingredient. A **double-blind** experiment also removes bias on the part of the doctors or researchers. A doctor in a study may administer a drug and measure the patient's response, but he doesn't actually know whether the drug or placebo was given.

Remember the Russian psychic who claimed to be able to manipulate brain activity and blood pressure? The investigator sent by the program *NOVA* designed a double-blind experiment to eliminate bias. The scientists and patient were kept in one room so his blood pressure and brain waves could be monitored. The psychic was placed in a separate room. The psychic randomly chose slips of paper that indicated whether the patient's blood pressure or brain activity would be manipulated, or if nothing would occur at all.

The scientists' readings on the patient matched the expected result by the psychic only one time out of four. This one match was likely due to random chance, and the experiment thus yielded no evidence that he actually possessed the claimed psychic abilities.

# Building Blocks of Nature

Environmental science is an interdisciplinary field of study. Aspects of fields as diverse as chemistry and sociology are needed in tandem to fully understand our relationship with the environment. Most of the scientific concepts in this field stem from biology, the study of life. This is natural, as a heavy emphasis is placed on the living parts of ecosystems—animals, plants, and microscopic organisms—and how they interact with each other.

However, a few other ideas that are more commonly associated with chemistry and physics are also important to have as background knowledge going forward. Is it possible to understand the effects of a pollutant without knowing its chemical composition? Acid rain and smog actually aren't directly released into the atmosphere; they're the products of a chemical reaction in the air! Laws of physics apply strongly to the production and use of energy, whether it be electricity from coal or gasoline from oil.

## Conservation of Matter

Matter cannot be created or destroyed; it only changes form. This is a short, simple law of nature, but an incredibly important one for multiple concepts within this book. When a chemical reaction occurs, essentially all of the matter that's present before the reaction is still present afterward.

When a few logs of wood are placed into a fireplace and completely burned, only a small amount of ash remains. This ash has much less volume and mass than the original logs did. What happened to the matter? The answer lies in the smoke that escaped through the chimney. While actual gases released when wood is burned are light enough to float in air, the total mass of all the compounds that left the fireplace accounts for what's missing from the ash. Ever wonder how it's possible that one person could be responsible for tons of carbon dioxide pollution over the course of a year?

 **CASE STUDY**

How do massive trees become so heavy? Where does all their mass come from? When asked, most people give answers that relate to the soil—the roots soaking up nutrients and other minerals from down below. A scientist named Jean Baptiste van Helmont decided to test this idea by weighing a willow tree seedling and the soil it would be potted in. He grew the tree for five years, measured it again, and found it had gained 164 pounds. He assumed the weight gain was completely due to water. Even this answer was not completely accurate, however, because plants are obviously not 100 percent water. The rest of the plant's mass actually came from carbon dioxide absorbed from the air through photosynthesis—the same carbon dioxide released when the logs are burned!

## Atoms, Elements, and Compounds

All matter in the universe is made from *atoms*. An atom is a very small particle that still contains all the original properties of the *element* it was taken from. If the atom were to be split any further, its properties would change.

Every atom is made from a combination of three smaller particles: protons, neutrons, and electrons. Protons are positively charged particles, neutrons carry no charge, and electrons are negatively charged. Look at the structure of a carbon atom:

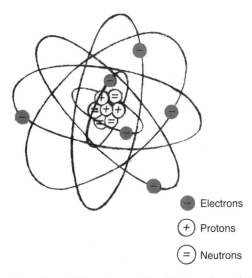

*In the carbon atom, the nearly weightless electrons (–) orbit the nucleus, which contains heavier protons (+) and neutrons.*

The nucleus contains six neutrons and six protons. The number of protons actually determines the identity of the atom. Seven protons would make a nitrogen atom, eight would be oxygen, and so on. Surrounding the nucleus is a cloud of six electrons. These electrons can interact with those of other atoms, creating larger *compounds* and *molecules*.

Molecules are particles made of more than one atom. The major constituents of air—nitrogen ($N_2$), oxygen ($O_2$), and carbon dioxide ($CO_2$)—are all molecules. Some molecules are made of only a single type of element, while others are combinations of two or more. These are compounds.

 **DEFINITION**

The **atom** is the smallest part of an element that still retains all the properties of that element. **Molecules** are made of two or more atoms joined together. When a substance is made of only one type of atom, it's an **element.** When it's made of two or more, it's a **compound.**

## Acids and Bases

Acids and bases are special types of compounds that influence an important property of water-based solutions: pH. What exactly does pH measure?

To answer that question, an understanding of water is needed. Water is a compound made of one oxygen atom and two hydrogen atoms. In any given sample of water, some of these molecules will break apart into two ions (atoms with a charge): $H+$ and $OH-$.

Some substances, such as hydrochloric acid (HCl), create hydrogen ions at a much greater rate. These lower the pH of a solution, making it acidic. Other substances, such as sodium hydroxide (NaOH), are able to actually combine with and absorb hydrogen ions. These raise the pH of a solution, making it more *alkaline* or *basic.* As seen in the following figure, the pH scale goes from 0 to 14. Directly in the middle, with a value of 7, is a perfectly neutral substance.

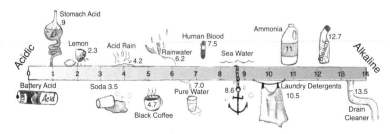

*The pH scale measures the concentration of hydrogen ions within a solution. Lower pH values are more strongly acidic, while higher pH values are more strongly alkaline.*

pH operates on a logarithmic scale, meaning that going up or down one number is actually a tenfold change. For example, you're comparing two carbonated beverages. The pH of the first one is 3.1, while the pH of the second is 4.1. This means the first beverage is 10 times more acidic. If you tested a sample of lemon juice and measured its pH as 2.1, then it would be $10 \times 10$, or 100 times more acidic!

pH is one of the most important physical properties of aqueous solutions, because it has major impacts on the life that can survive in it. As we'll cover later, most aquatic organisms have a limited pH range that's survivable.

## Energy

Two important laws of physics, called the laws of thermodynamics, are especially important to the study of environmental science.

The first law basically states that energy is never created or destroyed during a reaction or conversion; it simply takes on a new form. In other words, the quantity of energy going in and coming out of a system is the same. This might seem a little counterintuitive at first. After all, you fill up your gas tank, run your car for a week or so, and eventually the tank is empty. The car will stop running until it's refilled. According to the first law, the quantity of energy should be the same, right?

The second law of thermodynamics explains that whenever energy changes form, its quality decreases. As the car was driven, the high-quality energy of the gasoline was gradually converted into heat. Not intentionally, but this is a side effect of running a machine with many moving parts. The friction inside the engine, the brakes, and the wheels on the road all produce heat. That heat moves upward into the atmosphere and is no longer useful. The heat is considered low-quality energy.

As you'll see in different chapters later in this book, these two laws apply to all systems that use energy, both natural and mechanical. Living organisms produce body heat. This is the same kind of low-quality energy given off by the car!

Everything presented in this book is based on data and conclusions from researchers that all follow the scientific method. You may not often be given any 100 percent certain answers, but every concept is backed by peer-reviewed research.

## The Least You Need to Know

- Bias is the manipulation of a scientific study by people involved in the study—patients or researchers—to produce a false conclusion.
- All scientists design their experiments around the same basic method. This continuity ensures the most reliable results possible.
- In any chemical reaction, matter is never destroyed. It may be converted into a form that isn't easily perceived or measured, but it's still present.

- All matter is made of atoms—combinations of protons and neutrons in the nucleus surrounded by a cloud of electrons.
- The pH scale runs from 0 to 14. Highly acidic compounds fall on the low end, neutral compounds are in the middle, and alkaline compounds are on the high end.

# The Biosphere

Part 2 is all about life. We begin by exploring the vast diversity of life and resources present on Earth. You learn about the different types of species, how they evolved, why they look the way they do, and where they live.

Part 2 also covers some important concepts in ecology—the natural balance living things have with their environment. We look at the variables that influence population size and growth rate. Human populations will be ignored for the time being, since we can manipulate our environment more and don't always follow these same types of patterns.

We then explore each of the major types of ecosystems found on Earth, including on dry land, moving from forest to grassland to desert, and in the water, from the freshwater rivers and lakes to the vast oceans. The nonliving factors like climate and topography that influence each ecosystem are linked back to some of the biological and ecological concepts covered earlier. This part concludes with an overview of the movement of energy and matter needed to sustain these ecosystems.

# How Species Evolve and Interact

Ecology, the study of the environment itself, is no simple matter. A recent study estimates that 8.7 million different species reside on Earth. Each of those species interacts with other species in different ways, from hunting to competing for resources. There are also nonliving factors to consider, such as temperature, average precipitation, soil type, and so on. Any given ecosystem is a different combination of all these factors, making it unique.

In this chapter, we'll examine the living parts within eco-systems, called *biotic factors* or components. Biotic factors can include large, complex life forms like plants and animals as well as smaller, simpler ones like bacteria.

## In This Chapter

- Layers of organization within an ecosystem
- The biological system of classification
- How different species are defined
- The importance of natural selection
- Types of relationships species have with each other

# Ecology: Life and the Environment

The planet Earth is huge, with a pole-to-pole diameter of nearly 8,000 miles. However, only a small fraction of Earth, the part nearest the surface, actually contains life. This layer is called the *biosphere.*

Within the biosphere, multitudes of different *ecosystems* exist: tropical deserts, forests of evergreen trees, jungles, and frozen tundras. What separates these ecosystems? The answer is nonliving, or *abiotic,* factors. For example, tundras and deserts are both pretty dry—the average monthly precipitation on either is quite low. However, tundras obviously have a much colder average temperature. So what little precipitation does fall takes the form of snow and ice and remains there. Water falling in a desert, on the other hand, quickly evaporates or is absorbed into the soil.

Each ecosystem has its own unique collection of living things, all dependent on each other. This is the *community* within the ecosystem. For example, grassland may be home to flocks of sparrows and finches, herds of bison, and fields of, well … grass! Each group of species within an ecosystem is considered a single *population.* Finally, the fundamental building block of all this ecology is a single *organism,* like yourself!

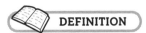 **DEFINITION**

> An **organism** is a single living thing. A group of organisms from the same species makes up a **population.** All the populations in a given area create a **community.** When you factor in the nonliving parts of the environment, you have an **ecosystem.** Finally, all the parts of Earth that support life make up the **biosphere.**

Another way of looking at the order of life on Earth is from the bottom up, from your own perspective. You are a single organism, but a part of the population of humans who occupy your immediate area.

Now, if you look outside, you will see and hear other living organisms—plants, birds, small mammals, and so on. You even have microscopic organisms living on and inside you right now. All together, these biotic factors make up your community.

Next, consider the climate where you live—the typical humidity levels, windiness, average precipitation, and presence of seasons. These are abiotic factors that define your ecosystem. Finally, if you found your location on the globe, you would see lots of other ecosystems all around you. You would be looking at the biosphere.

## Habitats and Ranges

When you consider most living organisms, they're found only in certain areas. Where would you go to find a polar bear? A chimpanzee? A giant cactus? The specific set of conditions, the ecosystem, that each of these species occupies is its *habitat*.

While knowing the habitat gives you a good idea of where an organism can be found, there's more to know. Cacti aren't found in every desert. There's a set of conditions that will limit where that cactus or any other organism can be found within the habitat. These are known as *limiting factors*. Common examples are temperature, presence of water, and soil type.

With each limiting factor, there is a "Goldilocks" effect—there can't be too much or too little for conditions to be optimal. As an example, consider the limiting factor of water temperature for fish living in a pond. If the water gets too warm, the fish will not be able to obtain enough oxygen. If the water gets too cold, their body metabolism will be affected. The fish species has a *range of tolerance* for temperature.

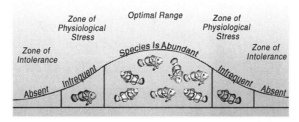

*This is a sample range of tolerance graph for the clownfish. Like other aquatic species, clownfish are very sensitive to changes in pH. Levels above or below the optimal range will have significant impacts on the population.*

The range of tolerance itself is divided into different sections based on how well the species is able to survive and reproduce. The optimal range in the middle represents the best conditions for the organism's growth. The population levels here are high. At the two zones of physiological stress, there's either too much or too little of the critical factor, decreasing the survival rate of the species. Finally, at the zones of intolerance, the levels of critical factor are so high or low that long-term survival is not possible. Population levels here are low to none.

## Physical Adaptations

How are ranges of tolerance established? Coniferous trees, ones that grow needles and don't shed their leaves, are found throughout North America. However, deciduous trees that shed their large, flat leaves during the winter aren't able to survive in the northern half of Canada. Why are some organisms able to survive over much wider ranges than others?

One of the key differences between related species that affects this range is the presence of physical adaptations. These are structural differences between species. These differences are extremely important in determining the organism's ability to survive in any given habitat.

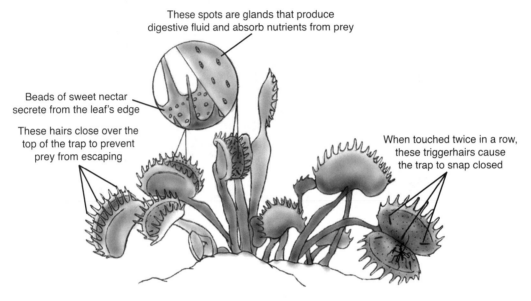

These spots are glands that produce digestive fluid and absorb nutrients from prey

Beads of sweet nectar secrete from the leaf's edge

These hairs close over the top of the trap to prevent prey from escaping

When touched twice in a row, these triggerhairs cause the trap to snap closed

*The Venus flytrap is a carnivorous plant with the ability to catch and digest insects. This gives it a source of nitrogen and other nutrients outside the soil and a big advantage over its competition.*

Primates are one group of related species with major differences in physical adaptations. How does the body structure of a gorilla compare to that of a capuchin monkey? Gorillas are much larger and do not have tails. Capuchin monkeys, on the other hand, are small, quick, and have prehensile tails they use to hold on to tree branches. Each of these primates lives within a jungle habitat, although on different continents. Gorillas live on the ground and get all their food from the forest floor. Capuchin monkeys spend most of their time in the treetops.

## Behavioral Adaptations

Physical and structural differences aren't the only way organisms can adapt to their habitat and extend their livable range.

Chimpanzees and bonobos are both primates, have very similar genetics, and live very close to each other. Their physical body structures are nearly identical. However, their social structure and behavior are worlds apart.

Chimpanzees live in large, male-dominated groups. Disputes are often settled through violent conflict and displays of physical dominance, and sex is strictly for reproductive purposes. Bonobos, on the other hand, are not male dominated, tend to be more socially peaceful, and engage in near constant sexual activity! The bonobos' unusual sexual habits, which include both heterosexual and homosexual acts, are believed to help strengthen the social bonds of the group.

*Violence and aggression are a way of life in chimpanzee troops. In particular, male-on-female violence has been observed throughout the species.*

What could possibly account for such wild differences in behavior between two otherwise nearly identical species? There must be a difference between the two ecosystems that favored the one behavior over the other. In other words, evolution occurred—nature selected for those behaviors.

# The Evolutionary Arms Race

According to the theory of evolution, all organisms originated from the same life form and are related. How, then, did millions of different species spring from one? Even given the billions of years that have passed since the first life forms inhabited Earth, this is still an extraordinary process to understand.

Every organism has physical adaptations, as described in the last section. These adaptations are not arbitrary; they exist because they increase the organism's chances of successfully surviving and reproducing. Keep this one simple idea in mind: every organism looks and acts the way it does for a reason.

 **CASE STUDY**

Why do male peacocks grow such heavy, elaborate tail feathers, reducing their ability to fly? Brightly colored feathers might seem like a disadvantage for the males, but they provide vital information for the female birds. After all, if a male is able to survive to adulthood and is healthy enough to develop such a colorful display, it must have a good set of genetic traits! These traits will be passed along to the chicks, furthering the species' evolution.

The reason, the advantage, is not always clear. Organisms don't exist within an ecosystem by themselves. There are many others, some of them possible food sources, others predators, still others parasites, and the rest inconsequential. With that in mind, how do you go from a single-celled bacterium to a giraffe?

## Diversity Within a Species

Take any population of organisms. Are they all identical clones of each other? Of course not. Just as an example, humans have a huge number of differences: eye color, hair color, skin tone, height, and body hair, to name just a few.

These differences are rooted in the DNA of each individual. DNA is a molecule that exists within every cell. It contains the instructions needed to build each cell and for the cells to perform their tasks. This includes physical and behavioral traits. When two organisms mate and reproduce, they pass along these traits to their offspring.

Occasionally, the DNA within a living cell will mutate. This mutation might simply result in a defective cell, but sometimes it creates a new variety of a trait. What will happen if this trait gives the organism an advantage?

## Emergence of a New Species

Consider a population of deer living in a coniferous forest in Northern Canada. Due to a mutation, some of the deer have slightly thicker fur than others. Imagine that these deer thrive in their environment and reproduce successfully for many generations. After a while, they'll start to experience a shortage in a critical factor, such as food. Part of the herd may splinter off and travel farther north, finding new territory with less competition for food.

The drawback to this migration by the deer is that they encounter another critical factor, temperature. The climate farther north has somewhat colder average temperatures. In this ecosystem, the deer with the thicker coats are more likely to survive in the long term. They retain heat better, conserving their energy and decreasing the amount of food they need to find.

The change in temperature is creating an effect called selective pressure. Survival is now more difficult for the deer with coats that are less thick. Over time, members of the population without the gene for thick fur will survive less often. Less survival means less reproduction. Less reproduction means the gene is passed along to the next generation less often. Over time, the gene may disappear entirely. What's happening to the deer is called *natural selection.*

How does this relate to the larger process of evolution? Remember, a species is defined as a population that will mate and produce healthy, fertile offspring. When the two populations of deer separated, the genetic diversity of the northern group began to change. Over time, each population will continue to change in response to its own ecosystem. At some point, the changes in physical traits and behavior will become so great that the two populations will no longer breed. Speciation has occurred; they have separated into two new species.

**A LOOK BACK**

One of the most famous places Charles Darwin visited was the Galapagos Islands near South America. There he noted several types of finches with many similarities that seemed to be closely related. One of the biggest differences between them was their beaks. Some of the birds had large, strong beaks they used to crush seeds, and others had developed very narrow ones to catch insects. Darwin believed each species came from the same original ancestor finch but evolved on the islands in response to different food sources.

To be more specific, the example of the deer is actually called *divergent* speciation because one group split from another. *Convergent* speciation occurs when two unrelated species develop similar traits because they live in similar ecosystems.

*The praying mantis and mantis fly both have incredibly similar barbed forelimbs that serve the same purpose—to capture prey. However, they belong to different insect orders and aren't closely related.*

# Relationships Between Living Species

Selective pressure by the environment is one means of driving evolution and speciation. However, there's another important factor: the influences of other species also trying to survive in that same environment.

A relevant evolutionary hypothesis here is called the Red Queen Hypothesis. In Lewis Carroll's *Through the Looking-Glass,* Alice and the Red Queen are engaged in a race. Yet in spite of their constant running, they remain in the same place. In fact, the queen remarks, "… it takes all the running you can do, to keep in the same place." In many ways, evolution is also a race—a race between predators and prey, a race between competitors for the same resources. If a population were to stop evolving new traits, behaviors, and adaptations, they likely would be overtaken by another species and go extinct.

Evolution, viewed across an entire community of organisms, is often compared to an arms race. As predators, prey, and parasites all reproduce, new traits and variations on traits emerge and accumulate over time. Thomson's gazelles have evolved an incredibly fast running speed of 50 miles per hour. This is essential in the open plains of Africa, where few hiding places are available. Not coincidentally, lions, one of their predators, have a similar top speed. Meanwhile, mosquitoes and other insects have evolved sharp, piercing mouthparts to feed off the blood of both animals.

## Energy in Food Chains

Mapping all the interactions among all the populations that make up an ecosystem's living community is no simple task. Organisms might have multiple types of relationships with each other.

One of the most apparent relationships is predator-prey. Simply put, the predator kills and ingests the prey as a source of energy for itself. Lions and gazelles, spiders and flies, and even a caterpillar and a leaf fit within this definition.

An important point to remember here is that this relationship is all about energy. One of the most important rules of ecology is that *energy flows through an ecosystem.* Follow the flow of energy through these organisms:

- A field of grasses, wildflowers, and other plants absorb energy from the sun.

- Grasshoppers ingest leafy flesh from some of the plants.

- A few of the grasshoppers are caught and eaten by a larger praying mantis.

- The praying mantis is eaten by a starling, a type of black bird.

This is an example of a food chain. Within food chains, different trophic levels or positions exist. Producers build their own sugar molecules from outside energy sources, like sunlight. Primary consumers, like the grasshoppers, ingest the producers. Secondary consumers, like the praying mantis, ingest the primary consumers. The starling, at the top of this food chain, is a tertiary consumer.

## Food Webs

Food chains are simple to make and easy to understand, but no ecosystem is limited to only four organisms. Thousands of species of predators and prey might co-exist within just one ecosystem.

To view the movement of energy through the entire ecosystem, ecologists make a kind of map called a food web. Food webs are made by taking all the organisms within an ecosystem and drawing arrows to show the flow of energy, much like a food chain. What makes food webs different is the amount of overlap between species. Within a freshwater river, a certain species of fish might prey on multiple species of insects. At the same time, it will likely have lots of predators, such as other fish, predatory birds, or even bears.

The basic components of the food web are the same as the food chain. All the trophic levels are present. Additionally, any decomposers that feed off dead organisms from any of the trophic levels are included. This provides the most complete picture possible of the energy movement within the ecosystem.

## Competition

While predator-prey is one of the most common relationships between populations in an ecosystem, it's not the only one. Different species are also constantly engaged in competition for similar resources.

If two animals share a common prey as their primary source of food, they'll compete for that prey. If two saplings grow close to each other in a forest, they'll compete for sunlight and water.

The most common type of competition is between different species. This is called *interspecific competition* and is a big driving force behind evolution. If one plant is able to grow faster or absorb more sunlight than another, it's more likely to survive and occupy that place in the ecosystem.

Competition also occurs between members of the same species. *Intraspecific competition* can be a positive thing in some cases, as when two males compete for access to a female for mating. After all, the species will be better off in the long run if only the strongest and fittest males mate.

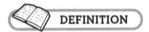

**DEFINITION**

There are two forms of competitive relationships in the wild. **Interspecific competition** is between two or more individuals of different species. **Intraspecific competition** is between members of the same species. The organisms may be competing over territory, food, or mates.

In other cases, intraspecific competition can be a detriment to the success of the species. Competing with other species for food and water is difficult enough, after all. Some organisms have evolved adaptations to avoid this.

Dandelion flowers produce seeds that can easily blow far away on the wind. This is known as dispersal, and it ensures that the parent will not have to compete with the offspring. Some animals, like wolves, establish and defend their own territory, which provides them with sufficient land to hunt food without competing with other wolves. Other animals, like insects, have different life stages that rely on separate food sources. For example, caterpillars typically eat leaves, while the adult butterflies rely on the nectar from the flowers.

## Symbiosis

Not all relationships involve predation or competition. In fact, some populations of organisms have close, intimate relationships with each other and would not be able to survive alone. These are called *symbiotic* relationships.

Each symbiotic relationship is named based on the dynamic that exists between the two populations. Aphids are small, soft-bodied insects that feed directly off the inner juices of plants. They aren't venomous and have no stingers to defend themselves. However, they do produce a sweet-tasting nectar that attracts ants. The ants feed off this nectar and in return defend the aphids from predators like ladybugs. In this dynamic, both populations benefit. This is called *mutualism*.

Not every relationship is equally beneficial. Whales are often covered with small, crusty barnacles. The barnacles don't harm the whale in any way (unless they grow over its blowhole), but they do take advantage of the whale's tremendous mobility in the ocean. This is an example of *commensalism*.

The last type of relationship has a much more sinister dynamic. Mistletoe, the plant we like to hang up during the Christmas holidays, is actually found on the branches of spruce trees. It roots into the vascular tissue of the tree, sucking all the water and nutrients it needs to live. Certainly, the tree (the host) is harmed in this relationship. This is an example of *parasitism*.

 **DEFINITION**

> **Symbiosis** refers to close relationships between two species. **Mutualism** is when both species in the relationship benefit. **Commensalism** is when one species benefits and the other is unaffected. If one species benefits at the expense of the other, it's called **parasitism.**

So how is parasitism any different from predator-prey? After all, one species is essentially consuming the other as a source of energy. The difference is, parasites usually don't actually kill the host—this would ultimately lead to their own demise as well. Instead, they extract just enough energy and nutrients to survive. This is why parasites are usually much smaller than the host.

## Matter Cycles, Energy Flows

These relationships between all the different populations within an ecosystem intertwine. Two species of plants compete for a bit of sunlight that penetrates to the forest floor. A primary consumer ingests one of those plants and then is hunted and killed by a secondary consumer. The primary consumer had a parasitic infection in its bloodstream, which is passed along to the predator when eaten. The secondary consumer, weakened by the parasite, is preyed upon by a tertiary consumer. The scraps left over from the kill are ingested by bacteria, fungi, and worms, which leave their waste behind. The waste becomes part of the topsoil, supporting another new sapling.

Within this ecosystem, all of the matter is used and reused constantly. This fits the law of conservation of matter, as none of it was created and none is destroyed. The energy, however, is constantly being converted and much of it lost as heat. Thus, the ecosystem requires a constant input of new energy. See why the sun is so important? Simply put, matter *cycles*, while energy *flows*.

## The Least You Need to Know

- A species' habitat describes what type of ecosystem it can survive in, while its niche describes its role.
- The behavioral and physical adaptations found within different species are the result of natural selection—the favoring of specific traits.
- New species emerge when a population splits into two that remain separated, and each eventually develops its own separate set of characteristics.
- Food webs and food chains show the flow of energy, starting with producers absorbing sunlight and ending with tertiary consumers or decomposers.
- Symbiotic relationships are the closest and most intimate and include mutualism, commensalism, and parasitism.

# Population Dynamics in Nature

Populations are groups of members of the same species living in an area. The actual size of each population varies. What determines the typical number of any given organism within an ecosystem?

Under normal circumstances, populations don't change drastically. Certainly they fluctuate up and down, but in general they tend to be pretty stable. However, there are a number of disruptions that can upset this balance, and many different types of growth patterns can happen as a result.

How would an ecosystem be affected if a predator was completely removed? Would the population of prey animals increase dramatically? Would another predator take its place? What influence would this have on the producer organisms farther down the food chain? The answers to all these questions are found within the study of population dynamics—how changes within an ecosystem affect the numbers of organisms within it.

## In This Chapter

- Factors that influence the size of a population
- The different types of population growth patterns
- Environmental limits to growth rates
- The consequences of rapid population growth

# How Does Your Population Grow?

Before we can look at how populations respond to a change, we need a baseline. How is the natural equilibrium established? Think about the four trophic levels of food chains: producers, primary consumers, secondary consumers, and tertiary consumers.

If you were to take a walk through a local forest preserve or park, what would you expect to see the most of? Producers, of course—green plants are everywhere! If you looked a little more closely, you could probably find plenty of signs of primary consumers, perhaps a chirping cricket or deer tracks. Secondary consumers might be a little more difficult to find, but there are probably a few birds or spiders around. However, you might not see any hawks, wolves, foxes, or other tertiary consumers. Why is this?

Remember the second law of thermodynamics: Energy decreases in quality during every conversion. Every stage of the food chain represents a conversion of energy. This means that some of the high-quality energy within a plant will be lost as low-quality heat when it's eaten and digested by an animal. How much? It varies, but a good estimate is based on the 10 percent rule. If you took a pound of corn and fed it to a group of animals, only about 10 percent of those calories would remain. This means that only a small fraction of that energy would be available for those at the top of the food chain! The remaining 90 percent is lost as waste heat that escapes into the atmosphere.

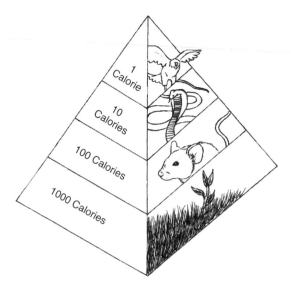

*Each level of the ecological pyramid is proportional to the population size of that trophic level.*
*Producers have the most individuals, while tertiary consumers have the least.*

The population of any given organism within an ecosystem is the result of the amount of energy available. Plants compete for the limited amount of sunlight that reaches the ground. Primary consumers compete for the most nutritious plants to eat. Secondary and tertiary consumers compete for other prey animals. Parasites try to steal energy from organisms in any of these trophic levels. Decomposers extract any energy remaining after an organism dies.

## Population Growth Variables

There are four different variables that will affect how a population grows or shrinks over time. Births cause a population to increase, as does *immigration,* or the movement of organisms into the ecosystem. Deaths and *emigration* have the opposite effect of shrinking the population. Given these factors, a simple mathematical equation for tracking population is:

$$\text{Change in Population} = (\text{Births} + \text{Immigration}) - (\text{Deaths} + \text{Emigration})$$

If *birth rate* and immigration are equal to *death rate* and emigration, the population is considered at equilibrium and doesn't change. However, this is pretty unusual. Populations are usually increasing or decreasing, at least a little.

 **DEFINITION**

> Four variables influence population size. The **birth rate** is the number of new individuals born into a population. The **death rate** is the number that perish over the same given period of time. **Immigration** is movement into an area, and **emigration** is movement out.

In order to track how quickly a population is changing, scientists create population growth curves. These are line graphs of population over time. Time is always on the x-axis, as it's considered an independent variable. Population size, being dependent on time, will be on the y-axis.

As an example, let's imagine that a population of 30 turkeys is introduced into a new ecosystem. These are specially bred turkeys that produce only two new chicks within the flock every year. The chicks always survive to adulthood. This is what the growth curve would look like:

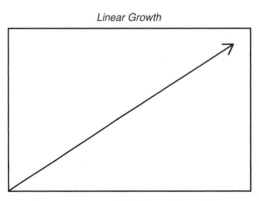

*Linear growth patterns are a straight diagonal line with a constant slope.*

This scenario and graph obviously are not very realistic at all. No population of living organisms restricts itself to a consistent level of growth like this. Additionally, a host of complicating factors is ignored. Do the turkeys have predators? Do any die of old age? What if there was a food shortage caused by a drought or blizzard? Do any diseases or parasites affect the population?

## Growing at Full Speed

Hypothetically, what would be the maximum population growth possible if every imaginable limit were removed? What if there were no predators, diseases, or limits on reproduction?

A famous tale that illustrates this kind of growth is called "The Sultan and the Chessboard." A skilled artist showed a sultan an intricately carved chessboard. The sultan was so taken with the beautiful chessboard that he asked the artist to name his price. The artist offered to sell the chessboard if the sultan were to give him one grain of rice, then double that amount for every square on the board.

At first this seemed like a bargain to the sultan. One grain, two grains, four grains, eight grains of rice. The amount kept growing by powers of two. By the twelfth square, the artist would receive about two handfuls of rice, then a pound of rice for the sixteenth square. However, by the twenty-fourth square, the artist would receive over one thousand pounds of rice. By the time he would reach the end of the chessboard, the sultan would owe more rice than had ever been produced in the history of civilization!

This kind of growth is called exponential growth, because it occurs in powers of two or another number. It starts off slow, but eventually the rate of growth is off the charts. This kind of growth seems unlikely for a living population of organisms, but it can happen—to a point.

There are a few occasions when organisms can reach their biotic potential and grow their populations at the maximum possible rate. This kind of growth makes the assumption that all the normal environmental limiting factors on growth are not present.

 **A LOOK BACK**

In 1944, the U.S. Coast Guard established a long-range radio navigation station on St. Matthew Island, 300 kilometers west of Alaska. The island was staffed by 19 men. Twenty-nine reindeer were brought to the island to serve as a backup food source in case supplies could not reach the men during periods of bad weather. When the station was decommissioned and abandoned soon afterward, the reindeer were left behind. This island was rich in their primary food source, lichen, and had no natural predators. The population grew to an estimated 6,000 over 19 years. By that time, the reindeer were no longer able to eat lichen, as it had been stripped from the island. As the other food sources disappeared, so did their population. Three years later, there were only 42 reindeer left. By the 1980s, all were gone.

A good example of this growth happens on a microscopic scale. Take a clean, sterile Petri dish and pour in a freshly prepared batch of agar. For certain types of bacteria, agar is the perfect food. Now, place one single bacterium on this dish, all by itself, surrounded by this massive (relatively speaking) amount of space and food. The bacteria will asexually divide into two. Those two bacteria will divide, leaving four. Those four divide into eight. Does this sound familiar? Each new generation of bacteria represents a different power of two. Two squared is four. Two to the third is eight. Let's say that the bacteria are able to successfully reproduce for 20 generations. How many would we have?

$$2^{20} = 1,048,576 \text{ bacteria!}$$

Ever wonder how those Petri dishes can become completely covered in growth after only a day or two? Because bacteria are single celled and essentially clone themselves, they can reproduce very quickly, and this process occurs at a tremendous speed.

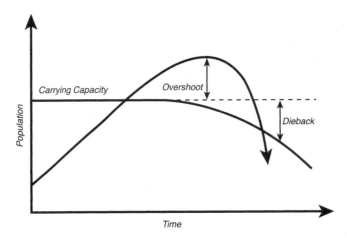

*Exponential growth is rapid, and can lead to a population exceeding the carrying capacity of an ecosystem. Disease or food shortage invariably leads to a sharp decline in the population.*

This kind of *exponential growth* is going to be unsustainable. The Petri dish is only so large, and the growing colony of bacteria will eventually begin pressing up against the edge. Open space will disappear, and the agar will be consumed quickly, replaced by the (likely toxic) waste products of the bacteria.

Faced with all these problems, the population will no longer be able to grow. It has reached the maximum amount the environment of the Petri dish can support, its *carrying capacity*. If this point is surpassed, it's referred to as an *overshoot*. The population will actually shift into reverse, falling at much the same rate at which it grew. This is called a *dieback,* or population crash. Many of the bacterial cells will die out, unable to absorb enough nutrients to support themselves.

 **DEFINITION**

> **Exponential growth** occurs when a population's initial growth rate is very slow but then begins to increase rapidly. The slope follows an exponential pattern ($2^2$, $2^3$, $2^4$, $2^5$, ...). Every ecosystem has a **carrying capacity,** the maximum population size that can be sustained. When this is surpassed, it is called an **overshoot.** The population will experience a rapid death rate, called a **dieback.**

# The Environment Resists Growth

Populations usually don't grow in this way. Food webs are too complex, and there are too many community interactions going on to allow an organism to reach its biotic potential.

The growth pattern most commonly seen in healthy ecosystems is called *sigmoidal* growth. Instead of the population explosion seen in the J-curve of exponential growth patterns, sigmoidal growth rates are slowed. A variety of factors begin to influence the population, interfering with its rates of reproduction.

## Growing to a Balance

Sigmoidal growth starts off slow and gradually picks up speed, just like exponential growth. However, as the numbers of organisms increase, the environment begins to put on the brakes. This is called environmental resistance, and it causes the rate of growth to slow more and more, until it eventually stabilizes completely.

Not coincidentally, the point at which the population size stabilizes in sigmoidal growth is below the carrying capacity line. While exponential growth curves overshoot this line and experience a severe dieback as a result, sigmoidal growth curves do not. The end point, instead, is equilibrium. What causes this environmental resistance? What factors interfere with the reproduction rates of these organisms so much that their upward growth turns into zero growth?

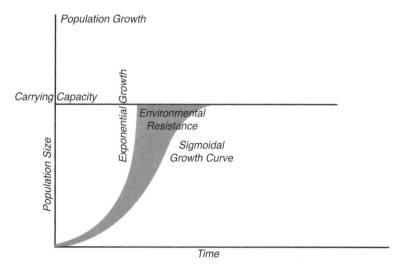

*Sigmoidal growth is often referred to as an J-curve because of the shape its graph takes.*

## The Challenges of a Crowded Population

One of the most common animals found on the floor of rainforests is actually an insect—ants. Ants are scavengers, eating dead or living organisms, depending on the species and what's available. They thrive in rainforests because of the sheer amount of life above them; plenty of nutrients and water are available. Yet, in spite of the abundant resources, ant populations will seldom experience exponential growth—they will rarely reach their maximum biotic potential.

When a population of ants starts to grow, some different forms of environmental resistance begin to curtail their reproduction and survival rates. As these are most likely to affect large, dense populations, they are called density-dependent limits to growth.

Ants can contract and spread disease, just like humans or any other organism. One of the most striking infectious agents to afflict ants is a fungus called *Cordyceps unilateralis*. The Cordyceps is a parasitic fungus that enters the tissue of its host and begins to grow. It actually affects ants' behavior as its filaments infiltrate their bodies. The ants become disoriented and begin climbing plants, grabbing the top with their mandibles before finally perishing. The fungus continues to grow and actually erupts out of the insect's exoskeleton. The body of the fungus will release thousands of tiny spores, some of which will land on other ants, continuing the cycle.

Cordyceps doesn't appear spontaneously; the only way a colony of ants could be infected is if a spore reached them from another infected ant. In other words, the fungus will spread much more readily and quickly if a lot of ants are living close together in the same area. Cordyceps, like any other disease, most severely affects dense populations.

Disease isn't the only density-dependent population growth factor. As the size of any population of organism grows, the amount of competition it faces increases as well. Tribes of chimpanzees, for example, have been observed to stalk and attack neighboring tribes in order to gain control over more territory and increase their own food source. These kinds of conflicts occur only when the population of one or more of the tribes grows so much that its current territory can no longer provide enough food.

These kinds of factors are most likely to affect populations in ecosystems with a lot of biodiversity. With so many different species present, relationships like parasitism, predator-prey, and competition are much more likely to be present. How, then, are populations limited in ecosystems with much smaller communities with fewer interactions?

## Limits That Affect All Populations

There are a few places on Earth that are so incredibly harsh, few species can successfully inhabit them. The frozen tundras of Northern Canada and Russia certainly fit that description, as do the hot, sandy deserts of Africa.

Species inhabiting these ecosystems are much more at the mercy of the physical, abiotic factors of their ecosystem, rather than of the biotic interactions of the community.

**CASE STUDY**

The best explanation for the Cretaceous-Paleogene extinction (when the dinosaurs died out) revolves around a giant asteroid strike. The asteroid would have caused large amounts of rock and dust to enter the atmosphere, blocking enough sunlight to inhibit photosynthesis in both land and aquatic ecosystems. Ecosystems that previously supported large populations of huge lizards now had a much smaller base of producers. This lowered the carrying capacity for every trophic level on the food chain, causing a population crash for nearly every species, especially very large ones like dinosaurs.

When you think of the Alaskan tundra, what comes to mind? A bleak, snow-covered landscape with no sign of life to be found? While this ecosystem does have a very long and bitterly cold winter, there is a summer, and it does support plants! When the spring thaw hits, many dormant species of grasses and wildflowers will suddenly spring to life. The landscape suddenly looks much more like prairie, at least for a few months. Animals will leave their wintering spots and take advantage of the sudden abundance of food. This growing season is short, however, and the freezing cold will again curtail the growth rates of all these species.

The changing climate, in this case, would be a density-independent limiting factor. The population of each species will grow and shrink in response to the onset of summer and winter. These impacts will affect all populations, regardless of how densely they have grown.

A good way of differentiating between density-dependent and independent limiting factors is to identify what's causing them. Density-dependent factors, as explained in the previous section, completely revolve around interactions with other organisms, from predators to parasites. Density-independent factors, on the other hand, are the result of changes in the physical environment: temperature, precipitation, sunlight, natural disasters, and so on.

Populations in ecosystems tend to form an equilibrium—a sort of balance between the nonliving factors in the environment and the other species living there. When any one population gets too large, a limiting factor pulls it back into balance.

## The Least You Need to Know

- Every ecosystem has a much greater number of producers than any other trophic level, because only a fraction of the energy in any given level makes it to the next level.
- Births and immigration cause a population to increase; deaths and emigration cause a population to decrease.

- Exponential growth is unusual in a biological community, and when it does occur, the population will eventually crash, due to disease or a lack of resources.

- Sigmoidal growth is much slower and more gradual, as a population experiences density-dependent limiting factors such as the presence of parasites and predators.

- Events that cause major changes to ecosystems, such as a natural disaster or climate change, are density-independent and affect all populations.

# Biomes: Ecosystems on Land

Think back to the different levels of the organization of life. What was the biggest and broadest level? The biosphere—the part of the Earth that supports life. Scattered throughout the biosphere are all kinds of ecosystems, from deserts to wetlands, coral reefs to frozen tundra.

What determines this distribution? Why is much of Northern Africa covered in desert, while the central part enjoys enough precipitation to support a jungle? Why does the city of London, England, only occasionally see snowfall, while a similar latitude in Quebec, Canada, is regularly buried in snow throughout the winter season?

The Earth's climate system is a complex one, with multiple factors and variables coming together to form different patterns of temperature and precipitation. These all together create the wide variety of land ecosystems, often called biomes, found across the Earth's continents.

## In This Chapter

- Geographic factors that most influence climates
- The types, formation, and life of deserts
- The distribution and diversity of grasslands
- Layering found within various forests
- How ecosystems change over time

# What Makes a Biome?

Ecosystems are primarily determined by climate. The most important factors are the average temperature and precipitation levels throughout the course of a year. On the dry or arid side of the scale are deserts and tundras. Deserts are hot and tundras are frozen most of the year, but each receives relatively little precipitation. On the other end of the scale are rainforests, which receive a consistently high amount of precipitation all year, supporting more plant growth than any other *biome*.

## What Determines Temperature?

The single biggest factor affecting an area's climate is geographic location. *Latitude* is a direct measurement of this variable. Latitude is given in degrees north or degrees south of the equator. In general, the farther away from the equator you go, the colder the average temperature of the biome. This should seem logical—after all, aren't all the frozen climates of the world found around the North and South Poles?

Certainly, latitude is not the only factor. Temperature can also vary greatly depending on whether the biome is close to a large body of water or farther inland. Consider, for example, the city of San Diego. The climate of this city is coveted by many in the United States, as a daily high temperature between 65° and 75° Fahrenheit is common. San Diego is directly on the Pacific coast, and biomes in this situation often have more consistent temperatures throughout the year. Farther inland, temperatures tend to vary much more. Staying within the United States, the average high temperature in the Dakotas can swing from nearly 100°F in the summer to well below zero in the winter.

Why does this happen? Water has a higher heat capacity. In other words, more heat is needed to raise the temperature of water than air. While the air temperature in Chicago in July might be pushing the 100°F mark, the water temperature is likely closer to 70°F. So if the winds are blowing in across a lake or ocean, they will be cooler.

*Altitude* also influences average temperature. The higher you go above sea level, the lower the average temperature is. This explains why snow can persist at the tops of mountains throughout the warmth of spring and summer.

 **DEFINITION**

A **biome** is a category of land ecosystem with specific characteristics. Biomes are defined by **altitude,** the height above sea level, and by **latitude,** the degrees north or south of the equator. The equator is 0°, the North Pole would be 90°N, and the South Pole would be 90°S.

## What Determines Precipitation?

Temperature is one big influence on biomes, but what about precipitation? Patterns of rainfall are a little more difficult to generalize, but there is one geographic feature that has a big impact: mountains. The way precipitation generally works is that water evaporates from soil, plants, or a body of water, enters the atmosphere, and condenses and falls back down at some point later. This entire process may be carried out over hundreds of miles as a weather system works its way across a continent. However, the presence of high-elevation mountains can accelerate everything.

When wind currents carrying moist air hit a mountain range, they naturally move upward. As the air moves upward, it cools, as higher altitudes have lower temperatures. Cool air holds less moisture, so most of the moisture condenses and falls to the ground.

This is known as the *rainshadow effect,* because a greater amount of precipitation falls on the side of the mountain that faces the incoming moist air (the *windward slope*). The *leeward slope,* facing away from the incoming air, receives very little precipitation, as most of the original moisture present has already condensed out.

**DEFINITION**

The **rainshadow effect** is the impact a mountain range has on precipitation levels in the biomes surrounding it. The **leeward slope** of the mountain range is on the opposite side of the prevailing winds and receives less precipitation. The **windward slope** faces the prevailing winds and receives a greater proportion of the incoming precipitation.

To see what impact this has on local biomes, consider the following example. The western coast of Washington State and British Columbia in Canada has an unusual type of biome called a temperate rainforest. This biome receives a lot of rain over the course of the year, yet the temperatures are pretty mild and seasonal.

As moist air blows into the area from the Pacific Ocean, it quickly hits the Olympic Mountains, followed by the Cascade Mountains. As you travel up the peaks of these ranges, snow and ice are seen throughout the year. Down the eastern side of these slopes, much less precipitation falls. Relatively few trees grow here, replaced instead by grasses and cacti. Over a distance of just a few miles is a drastic shift in ecosystems, all due to the presence of mountains!

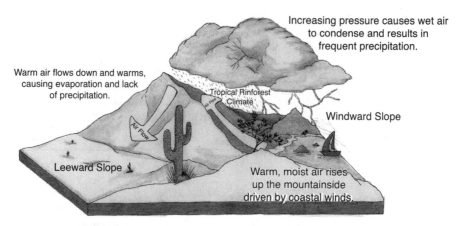

*As it rises up a mountain, moisture from the ocean condenses quickly due to the increasingly cold air.*
*Heavy rainfall results on the windward side, while the leeward side receives very little.*

## A Biome in One Graph

Given that biome type is most influenced by precipitation and temperature, it should be pretty easy to classify and compare ecosystems if this data is available and presented in the right way. This is where the climatogram comes in.

The climatogram is basically two graphs in one. It includes a line graph showing change in average monthly temperature over a year, and a bar graph showing the same for precipitation.

Climatograms are a quick and easy way to assess a biome—how much heat and rainfall it receives, and if either varies during the year.

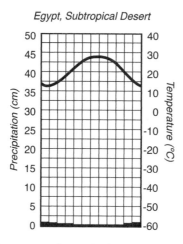

*Egypt's climate is hot in the summer and warm in the winter and dry throughout the entire year.*

# The Dry Deserts

Deserts are the driest of all ecosystems. Over the course of a year, a desert may receive only a few inches of rainfall. The precipitation is infrequent and unpredictable. As a result, the overall amount of life is low compared to other biomes.

The lack of water is a big limiting factor on plants. Without plants serving as the basis of the ecological pyramid, few primary consumers can be supported and even fewer secondary and tertiary consumers.

When you think of a desert, the first image that comes to mind might be that of rolling, lifeless sand dunes, unrelenting heat, and not a sign of life to be seen for miles. While there certainly are deserts that fit this description, there is much more variety in their biodiversity and climate than you might think.

## Subtropical Deserts

Many deserts are located near the equator, between about 20° and 40° north or south. With this proximity to the equator, these deserts are consistently hot all year round. Very little variation in temperature occurs from month to month.

As the hottest and driest of all the deserts, subtropical deserts have the least amount of biomass, or life. Living things do exist here and must be adapted to survive in such dry conditions. For example, some plant seeds are able to lie dormant in the sands until water arrives. They immediately germinate, come into bloom, and produce seeds before the water again disappears.

Why are these deserts so dry? The answer is not a simple one, but as an example, let's consider the Atacama Desert in Chile. Next to Antarctica, the Atacama is the driest region on Earth, with some areas never receiving a drop of rain since record keeping began. The Atacama is directly adjacent to the Andes Mountains, and the prevailing wind currents in that part of South America go from east to west. Due to the rainshadow effect, the eastern windward side, which contains the Amazon rainforest, gets a much greater amount of precipitation. The western leeward side gets almost none, creating the desert.

## Temperate Deserts

Temperate deserts, by definition, fall in latitudes between the subtropics and subarctics. This means between 40° and 60° north and south of the equator. At this distance, the temperatures show much more variability.

Temperate deserts experience seasons primarily based on temperature changes. Often, at least one winter month will have an average temperature below freezing, while the summer months will consistently breech 100°F.

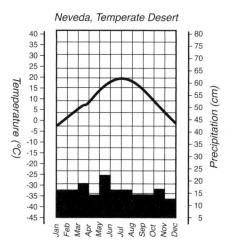

The climate of a temperate desert is marked by a greater degree of cooling in the winter months and overall higher precipitation than a subtropical desert. These deserts are still dry, although they do receive a bit more precipitation than those in the subtropics. Most of this precipitation falls in the cooler winter months, and is enough to support plants like giant cacti and sagebrush. Not enough precipitation falls to support large, tall trees.

## Polar Deserts

Polar deserts, also called ice or frozen deserts, are extremely dry and bitterly cold during the winter months. These are found only in polar and subpolar latitudes approaching 90 degrees.

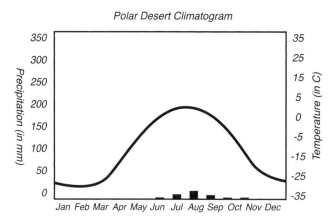

*The temperature in polar deserts drops enough in the winter months to cause the little precipitation that falls to freeze and take the form of snow.*

Most of the interior of Antarctica is considered a polar desert. This may seem counterintuitive; after all, isn't the land base of Antarctica covered by miles of sheet ice? While this is true, this ice has taken millennia to accumulate. The actual amount of rainfall that hits these areas on a yearly basis is extremely small.

**CASE STUDY**

The driest of the polar deserts is undoubtedly the McMurdo Dry Valleys, located in Antarctica. The surrounding mountains effectively block any flowing ice from reaching the valleys. This ecosystem has extremely low humidity and no snow or ice cover.

The dry valleys are not completely lifeless, as bacteria have been found living in protected areas inside rocks. Mummified seal carcasses have been discovered here, perfectly preserved in the dry air. The usual organisms that would decompose them, bacteria and fungi, are unable to survive in this incredibly harsh climate.

# The Rolling Grasslands

The single biggest limiting factor to life in the desert biomes was precipitation. Indeed, that trend persists into the grassland biomes as well. While they do receive more precipitation on average than deserts, the amount still dwarfs that of a forest. As a result, the varieties of plants and animals are still limited to an extent.

Grasslands exist in similar varieties as deserts: tropical, temperate, and polar. The seasonal temperature changes are vastly different in each of these biomes, and they have huge impacts on the characteristics of the grasslands.

## Tropical Grasslands

Tropical grasslands are named so because of their closeness to the equator. Because they exist at these latitudes, the average monthly temperature changes little from month to month. However, these ecosystems still experience seasons!

Not all seasons are defined by temperature changes. In the case of tropical grasslands, they experience wet and dry seasons. Seasonal shifts in prevailing wind patterns and ocean currents can direct heavy amounts of moisture into this biome during some months and arid more desertlike air during others.

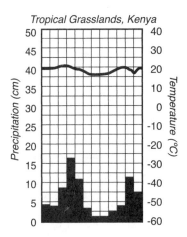

*Tropical grasslands, due to their closeness to the equator, show very little variation in temperature. Rather, they experience wet and dry seasons.*

The biggest continuous tropical grassland in the world is the African savanna. This biome starts at the very southern edge of the Sahara subtropical desert and wraps around the tropical rainforests of central Africa. The landscape is dotted with small, isolated trees and shrubs. The amount of rainfall received during the wet season is enough to keep these plants growing, and they then serve as a food source for the rest of the food web.

## Temperate Grasslands

Farther away from the equator are the temperate grasslands, also known as prairies. The temperature in this biome is much more variable than tropical grasslands. The summer temperature is hot and comparable to that near the equator. Spring and fall temperatures are much cooler and milder, and the winters can be as cold and harsh as those near the Arctic.

The seasons in temperate grasslands are defined completely by these changes in temperature. The precipitation levels are pretty consistent throughout the year.

One of the distinguishing features of temperate grasslands is the extraordinary depth of the upper nutrient-rich layer of soil, called topsoil.

The source of this exceptionally rich soil layer is twofold. Temperate grasslands often experience fires, especially during the hot summer months. As a result of these fires, minerals stored up in plant tissue are quickly recycled back into the soil. Additionally, as winter sets in, all the grasses and other plants die back on the surface, leaving more to decompose and add to the soil nutrients. Temperate grasslands, such as the prairies of the Midwestern United States, are considered some of the best biomes to use as cropland.

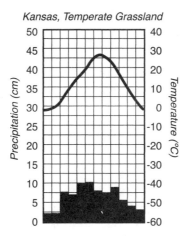

*Kansas, Temperate Grassland*

*Temperate grasslands are found mainly in the interior of continents and experience huge swings in temperature over the course of a year.*

 **A LOOK BACK**

When the first settlers moved into the Midwest in the eighteenth century, they primarily kept to the forested areas. The tallgrass prairie was seen as unfarmable, as the root mat of the plants there was incredibly thick and tough. This changed when the steel plow was invented by John Deere in the early nineteenth century. Farmers discovered a deep, rich layer of topsoil down below those plants, and the rapid conversion of prairie into farmland began.

## Polar Grasslands

Much like the subtropical deserts, many people assume the polar regions are completely devoid of life. While these biomes are very harsh ones, with freezing temperatures most of the year and relatively little total precipitation, life can still thrive there.

Polar grasslands, also known as Arctic tundras, are found in latitudes leading up to the North Pole. This type of ecosystem isn't found in Antarctica, as the land there is covered in such a thick sheet of ice that plants are unable to take root.

In spite of its proximity to the pole, the Arctic does have seasons. However, the winter season is much longer and colder than in temperate regions, severely limiting what life can survive here. Depending on how far north you go, the growing season may be only a few months long. But during this short time, the snow thaws, the soil is exposed, and plants are able to photosynthesize sunlight.

Polar grasslands, at least in summer, actually do resemble their temperate cousins. There are plenty of primary consumers that come out, animals like moose and caribou that gorge themselves on the suddenly abundant plant life. A few secondary and tertiary consumers appear as well, such as Arctic fox, Arctic wolves, and polar bears.

The limiting factor to life here is the drop in temperature during the winter months. The plants become dormant during this time, and the animals must either migrate to other areas in search of food or hibernate, or both.

# The Lush Forests

When enough precipitation is present and the growing season is long enough, large trees can be supported. Not the sparse, solitary trees seen throughout the African savanna, but deep, thick forests with so much growth that hardly any sunlight reaches through the upper layers of leaves and branches.

All types of forests are dominated by trees, but that's about the extent of what they have in common. There's a huge variety in size, shape, leaf type, and growth rate of the different species of trees found throughout the biosphere. Each thrives in a different set of conditions, creating three vastly diverse types of forest biomes.

## Tropical Rainforests

The forests closest to the equator have the fewest number of density-independent limiting factors—that is, factors primarily caused by the physical environment. In other words, they have abundant warmth and sunlight, as the equator does not experience a seasonal shift in temperature. The precipitation levels are also consistently high.

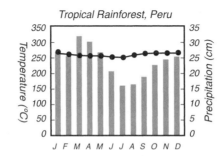

*Tropical rainforests show very little seasonal variation. The climate is warm and moist throughout the entire year.*

Having removed the limiting factors of sunlight and water, what prevents any given species of tree from experiencing a population explosion? The answer is competition. Rainforests have a greater amount of biomass and biodiversity than any other ecosystem, with the possible exception of wetlands. This includes not only the plant producers, but animals at all other trophic levels, such as primates, insects, and birds.

Given the tremendous amount of life cycling throughout this biome, you would expect the soil to be incredibly rich and full of nutrients, perhaps even surpassing that of the temperate grasslands. In reality, however, the soil is actually quite thin and nutrient-poor. Why? Remember, there is a tremendous number of plants, big and small, competing for any available space and sunlight, each looking for an advantage. Whenever any nutrients are recycled back into the soil of a rainforest, they are almost immediately taken back up by a fern, vine, or one of the species of hardwood trees nearby.

 **CASE STUDY**

The thin and nutrient-poor soil of the rainforest has led to a damaging agricultural practice known as slash-and-burn agriculture. Loggers go into an area of rainforest and remove any timber that can be sold. The remaining trees and plants are burned, leaving a layer of ash that is tilled into the soil as a source of nutrients. This creates a plot of fertile farmland, at least for a while. Eventually, the soil is completely depleted and no longer farmable. The people who worked the plot move on and repeat with a new area of forest. Meanwhile, the soil left behind is very prone to erosion and unable to sustain much plant growth at all.

## Deciduous Temperate Forests

Not all forests have the luxury of a yearlong growing season. Deciduous temperate forests, located in similar latitudes as the temperate grasslands, experience a wide range of temperatures from the freezing winter through the hot summer.

Cold winters present a huge challenge for hardwood trees. The air in winter is much drier than in summer. Broad, flat green leaves are great at absorbing sunlight needed for photosynthesis, but are prone to losing moisture. If these leaves were maintained throughout the winter, the entire tree would die of moisture loss. So as an adaptation, the tree stops producing chlorophyll in the autumn months, causing the leaves to change color and eventually fall to the ground. The tree sits dormant before regrowing the leaves in the warmth of spring.

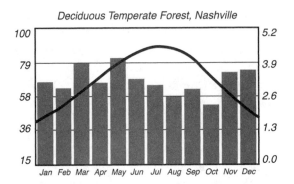

*Temperate deciduous forests experience enough consistent precipitation to support tree growth. The winter months are mostly cold and dry.*

Deciduous forests and tropical rainforests both have layers, with different living things present at each level. The canopy, or top layer, tends to hold most of the leaves and fruits of the trees and attracts a large number of animals. The understory, below the canopy, contains much fewer branches and leaves by comparison. Most of the sunlight is absorbed in the canopy, so producing leaves at this level would be a waste of energy for the large trees. Only a few saplings, desperately trying to soak up the remaining rays of sunlight, are found in this layer.

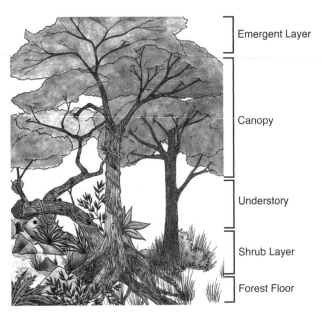

*The layers of a forest are defined as much by their height as by the type of growth found. From the thick canopy down to the decomposers of the forest floor, each is its own individual ecosystem.*

The shrub layer, toward the bottom of the forest, contains smaller plants such as ferns that produce disproportionately large leaves. Again, this is an attempt to absorb whatever sunlight remains at this level. Finally, the forest floor primarily contains dead and decomposing plant matter that falls from above. Most of the decomposers and scavengers of the forest—ants, fungi, and bacteria—reside here.

## Boreal Forests

Not all trees have the wide, flat leaves seen in the tropical and deciduous forest biomes. Forests in the subarctic regions are dominated by coniferous trees that have needles for leaves.

While needle leaves don't have the surface area or amount of chlorophyll their broadleaf brethren do, they bestow several important advantages to the trees in these frigid climates. First and foremost, these leaves don't need to be shed during the winter months. They have a waxy coating that prevents moisture loss. This is an important adaptation to the subarctic climate, because the growing season simply is not long enough to justify shedding and regrowing leaves. This gives coniferous trees the additional advantage of absorbing sunlight throughout winter, while deciduous trees are dormant.

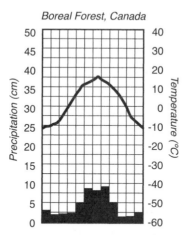

Boreal forests experience winter months that are longer, drier, and colder than deciduous temperate forests. This shortens the growing season, restricting what trees can survive. Coniferous trees grow into a much different shape, as well. If you were to take a moment to sketch a simple Christmas tree, what shape would it be? Triangular. This shape allows snow that builds up during the long winter to simply slide off the branches, instead of dangerously weighing them down.

The last big advantage of needles is in their chemical nature. They are acidic, and this serves as a deterrent for hungry animals during the winter. This adaptation is not perfect, however, as moose and a few other species are able to digest the needles.

# Ecosystems Age and Mature

There's obviously a tremendous amount of diversity in the different biomes of the world. In addition to the various types of grasslands, forests, and deserts we've covered, there are other, smaller biomes such as the dry chaparral of southern California.

Furthermore, when ecosystems change, one does not suddenly stop where the other begins. You won't necessarily see a clear border between a temperate grassland and a deciduous forest, for example. There may be miles of transition area where some trees are present, but not the numbers or thickness of a forest. These transition areas are called ecotones, and can be found in the adjoining areas of all biomes.

The final important feature of biomes is their changing nature. The Earth is not the same place it was millions of years ago. Continents have shifted, climates have changed, and the atmosphere has a different composition. Biomes are dynamic, and will be altered anytime one of these variables changes. This change is known as *succession*.

## Growing from Nothing

Succession occurs in predictable series of stages. The exact stages a biome goes through depends on what precipitated its succession. If we are talking about a brand-new ecosystem developing on a newly formed piece of land, this is primary succession.

How is it possible to have new land not yet colonized by life? Consider volcanic islands. An eruption releases large amounts of molten rock, which cools as it hits the seawater, building up more and more on the ocean floor. Eventually, enough cooled magma amasses to form an island. In its initial stage, no life exists on the island.

At this point, the island isn't ready to be colonized by large plants. There isn't really any soil for them to root into, just ash and rock! Instead, what grows first are smaller producers like lichens and mosses. Lichens are actually two organisms living in a mutualistic relationship: a fungus and a photosynthetic algae or bacteria. They have the ability to survive on rocks. Not only do they survive on the rocks, but they chemically degrade the rocks, breaking them down into the smaller minerals that make up soil. Organic material also may be introduced at this point, such as droppings from seabirds passing through.

Eventually, the rocks will have broken down and combined with enough organic material to form soil that can support small grasses, herbs, and shrubs. A few tree seedlings may even begin to sprout. This collection of life is called the pioneer community, because these organisms are able to survive and grow quickly in this newly developing ecosystem.

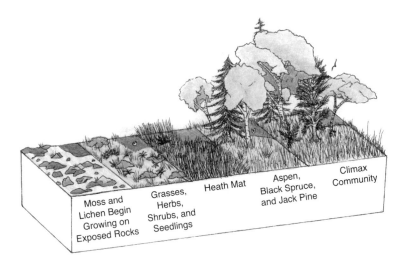

*Ecosystems that arise from the bare, lifeless rock of new land do so through the gradual process of primary succession.*

Over time, as the pioneer plants grow, die, decompose, and new ones regrow, the topsoil will get deeper and more nutrient-rich. The struggling seedlings of the pioneer community will grow into full-size hardwood trees. Eventually these early trees will die, as they tend to have a shorter life span. Trees like this, such as aspens and red oaks, will eventually be replaced by mature trees with a much longer life span, such as birches and white oaks. At this point, our ecosystem is considered in its final stage. This is called a climax community.

## Rising from the Ashes

What if the ecosystem is an already existing one, but something catastrophic happens to it, wiping out its entire living community? This is more common than you might think. Natural disasters like hurricanes, floods, fires, and volcanic eruptions can have this effect. It can also be caused by human activity, such as clear-cutting a forest. So what happens? Does the devastated ecosystem remain forever lifeless?

The answer is again succession. This type of succession is different, as it is occurring over an already existing ecosystem. This is called secondary succession.

The basic stages of secondary succession are the same as primary succession. You start with a group of pioneer species, which immediately colonizes the area and grows very quickly. As their life span is short, they are eventually replaced by larger organisms with a greater life span—the climax community.

The main difference is, since the ecosystem already existed, this type of succession doesn't take as long. Before, we started with rocks and had to wait for the rocks to erode due to wind, water, and the action of lichens. Here, that work has already been done. Soil is already in place. Within days or weeks of the disturbance, the first pioneers, weeds, will appear. It may still take centuries to return to a fully mature climax community, but this is a fraction of what it takes to reach the same point in primary succession.

The wide variety of ecosystems across the Earth are mainly the product of differences in temperature and precipitation. Lots of factors can influence these changes—latitude, altitude, and so on. Each species that resides within these ecosystems has carved out its own niche based on these factors.

## The Least You Need to Know

- Each biome is determined primarily by two climatic factors: average temperature and precipitation.
- Average temperature tends to decrease as latitude and altitude increase. Proximity to water results in milder temperatures.
- Subtropical deserts are the hottest and driest, followed by the more seasonal temperate deserts and the frozen polar deserts.
- Tropical rainforests receive the greatest amount of annual precipitation and have the longest growing season, and thus contain the most biodiversity.
- Deciduous trees shed their leaves in the winter as a way to avoid moisture loss in the dry, frigid air. Trees of the boreal forest have needles for leaves to preserve their moisture and discourage animals from eating them.
- Primary succession occurs when a new biome develops on land that had no life before. Secondary succession takes place in existing ecosystems that have suffered a disturbance.

# Aquatic Ecosystems

Water covers the majority of the Earth's surface, about 70 percent compared to land. The first life on Earth began in the water. We can trace the ancestry of each of the species of organisms alive today back to an aquatic ancestor. When scientists survey the universe seeking other planets that may contain life, they look for liquid water. Suffice to say, water is an extraordinarily important element on our planet and a key ingredient to life.

Aquatic ecosystems are those that are covered with water permanently, as is the case with lakes and oceans, or at least part of the year, as with wetlands. Diverse communities of living organisms are found just about any place on Earth where there is water, from the ice sheets of the poles to the hot springs at Yellowstone National Park.

## In This Chapter

- Factors that influence life in aquatic ecosystems
- Different zones within rivers and lakes
- The types of wetlands: bogs, swamps, and marshes
- Ecosystems with the most diversity
- Unusual food webs at the bottom of the sea

# Abiotic Factors in Water

At first glance, it would seem that aquatic ecosystems couldn't possibly have the vast diversity of the biomes seen on land. After all, one of the most defining limiting factors on land is precipitation. In aquatic ecosystems, this is generally not a factor—unless they dry up in a drought!

However, aquatic ecosystems have several other important differences that define what organisms can and cannot survive there.

Some aquatic ecosystems are relatively permanent. Assuming no catastrophic events occur, deep oceans, shallow seas, and large lakes will always have water. Other ecosystems are a little more dynamic. The depth and speed of rivers change during storms and dry periods. Wetlands can be completely submerged or dry enough to walk on. The actual amount of water present is an important factor for these ecosystems, but it is far from the only abiotic factor.

## Salinity and Sediments

The water found in any aquatic ecosystem is not pure water; a multitude of other substances are dissolved in it. This is important, because many of these substances are vital for sustaining life.

The biggest difference between oceans and lakes is the presence of salt. The oceans contain an enormous amount of salt, somewhere on the order of 50 million billion tons! A good way to visualize the difference between marine, or saltwater, ecosystems and freshwater ecosystems is to take a water sample from each and heat it. When heated to the boiling point, water changes phase, becoming a gas and evaporating. Meanwhile, the solids dissolved in the water have a much higher boiling point and will remain behind. So if you boiled 1 cubic foot of ocean water, more than 2 pounds of salt would be left behind. A similar amount of water from a freshwater lake would contain less than a single ounce!

What are salts? They're minerals, like sodium, potassium, and calcium. Many of the minerals in the ocean came from volcanic vents at the floor. The salt level of the ocean isn't consistent but varies, depending on the composition of the ocean floor and whether any rivers are flowing into the immediate area.

**CASE STUDY**

The Dead Sea is a deep salt lake bordering Jordan, Israel, and Palestine. It has a salinity level more than eight times greater than the ocean, making it one of the most concentrated sources of saltwater in the entire world. As a result, very few organisms can survive there, hence the name "Dead Sea." Water in cells has a tendency to follow higher salt concentrations, causing living cells to shrivel. Marine organisms have adapted to manage this problem, but only a few species of bacteria are able to survive in the Dead Sea. These are called *halophiles*, or salt lovers.

Salts are not the only dissolved solids in water. Due to the vast amount of life found in the various aquatic ecosystems, there are large amounts of elements from living matter—from dead and decomposing organisms or excreted waste and urine. Regardless of the source, the amounts and types of dissolved solids in the water play a big role in shaping the communities found within.

## Temperature and Dissolved Gases

One abiotic factor the aquatic ecosystems share with those on land is temperature differences. While water does not gain or lose heat as easily and quickly as air, there are still some dramatic differences in temperature throughout the world's bodies of water.

The ocean water nearest the poles is often near the freezing point. Indeed, during the winter months some of the water will freeze and become sea ice. Meanwhile, the waters near the equator are so warm that their energy can feed the growth of tropical storms and hurricanes.

Temperature also directly influences the levels of dissolved gases in water. Imagine that you open two bottles of carbonated soda—one from the refrigerator and one that has been sitting out at room temperature. Once you break the seal and open each bottle, dissolved carbon dioxide will begin to escape. Does it escape at the same rate from each bottle? Which would you expect to go flat first?

Warm liquids don't hold dissolved gases as well as cold liquids. How is this relevant to life in aquatic ecosystems? Just as on land, animals in water need to breathe oxygen. The difference is, they extract it from water instead of air. Warm bodies of water have lower levels of dissolved oxygen (DO). Temperature affects how well other gases dissolve, too, like carbon dioxide; that will come into play more in Chapter 20, on global warming.

## Depth and Sunlight

The last important property of water is its ability to absorb sunlight. This has a drastic effect on the food webs found at different depths of lakes and oceans.

The top layer of a lake or ocean has the greatest amount of sunlight penetration. As a result, photosynthetic organisms like algae and plankton thrive there. This layer is called the *euphotic* (Latin: "good light") *zone*.

The amount of sunlight diminishes significantly as you move deeper into the water. Consequently, the population of photosynthetic organisms drops. This area is called the *aphotic* (Latin: "without light") *zone*, although a few wavelengths of light may still penetrate into this layer, depending on the exact depth.

Finally, the dark physical bottom is the benthic zone. Depending on the depth of the body of water, this layer may be completely devoid of light. The community of organisms down here is vastly different, as photosynthesis is impossible.

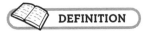

**DEFINITION**

> The two biggest differences between aquatic ecosystems are the levels of light and salt. Marine ecosystems contain saltwater, while freshwater ecosystems do not. Marine ecosystems are often deeper, so they're separated into three zones based on how much light penetrates through the water. The top layer, where photosynthesis can occur, is the **euphotic zone.** Some light penetrates through the **disphotic zone,** but not enough for photosynthesis. The lowest, completely dark layer is the **aphotic zone.**

# Freshwater Ecosystems

Freshwater ecosystems are those without a significant amount of dissolved salts. All freshwater ecosystems are found in the interior of continents. These ecosystems are generally fed by rainwater or melting ice, so their depth and temperature may vary seasonally.

Freshwater ecosystems tend to show much more variation in the characteristics of the water itself than those in the ocean. This is because these ecosystems tend to be much smaller and more easily influenced by changes in climate and the land that surrounds them.

## Rivers

Naturally, the best place to start with rivers is their source. The most common river source is actually the melting of frozen glaciers, such as those found at the top of large mountains. The major rivers of Asia, the Ganges and the Yangtze for example, are fed by the melting glaciers of the Himalayas. There are other sources, too. The single biggest source of water for the Mississippi River is outflow from Lake Itasca in Minnesota, which itself is fed by several tributaries—smaller rivers and streams.

The water at a river's source tends to be colder and lower in nutrients. This is because many rivers begin at glaciers or higher altitudes. The nutrient level is low because the water has not passed through enough land to absorb soil and other sediments that would serve as nutrients for plant life. The oxygen level also tends to be high, as colder water holds more dissolved gases.

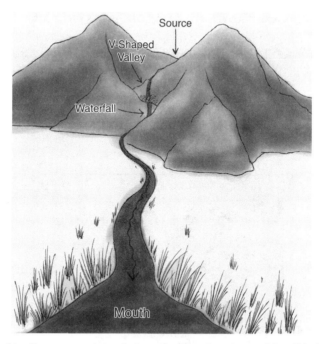

*Cold water melting from a mountaintop glacier and passing through waterfalls will be high in oxygen but low in nutrients. By the time the water exits the river's mouth, it's warmer, lower in oxygen, and much higher in nutrients.*

As a river winds through its course, it tends to slow. Its temperature will likely be warmer, as it has been exposed to more sunlight. The nutrient level of the water also begins to increase.

Eventually, all river systems end, flowing into either another river, a lake, or an ocean. This junction is called the river's mouth. By this point, the river has absorbed the greatest amount of sediment from land, creating a plume that can be seen from above. If the river empties into the ocean, it's called a delta—and this ecosystem has its own unique characteristics, which we will cover in the "Coastal Ecosystems" section.

## Lakes and Ponds

Lakes and ponds are inland bodies of standing water. There isn't really any official distinction between the two, but lakes are generally larger and ponds smaller in size. They do share a lot of common features, including a shallower area near the shoreline and a deeper, colder area toward the center.

Perhaps the most biologically active part of lakes and ponds is the littoral zone. This is a shallow zone along the outer edge of the water. Sunlight is able to penetrate through this entire zone due to the shallow water. This, combined with the heavy amount of dissolved sediment from the shore, allows plants to thrive.

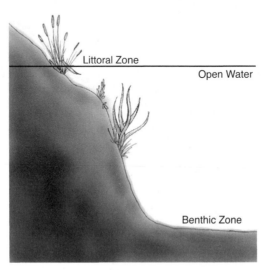

Littoral Zone

Open Water

Benthic Zone

*The littoral zone of a lake is the shallowest, warmest, and most likely to contain producers like plants and algae. The benthic zone is darker, colder, and higher in decomposers.*

Away from the shore of the lake or pond, the water becomes deeper and cooler. There's an area of open water toward the middle, but this tends to be the least biologically active of all the zones due to the distance from land.

Far below the surface, at the bottom, is the benthic zone. Depending on the water depth at this point, little or no sunlight is available for photosynthesis. Instead, the food chain in this zone is filled with decomposers, filter-feeders, and bottom-feeders. What makes the basis of this food web, if sunlight is not available? Basically, anything that falls down from above: dead organisms or waste.

**CASE STUDY**

Before the twentieth century, the Salton Sea was a dry lake bed near the Colorado River in California. In 1900, irrigation canals were dug from the Colorado to this area, creating fertile land that could be farmed. After a few years, the canals became blocked as sediment built up from the water flowing through. A wet spring in 1905 caused the river to swell, overflowing the dikes and completely flooding the area. The dry lake bed became an actual lake, which remains there today. This accidental flooding spurred construction of the Hoover Dam, farther upstream in the Colorado.

# Freshwater Wetlands

The last category of freshwater ecosystem is also the least permanent. Wetlands are technically defined as areas that are underwater at least part of the year. They also contain special kinds of plants called hydrophytes. These plants can survive in soil that has limited oxygen availability. This is important, because wetland soils are so saturated that air can't always penetrate to the roots of the plants.

There are three main categories of wetlands, each defined by the diversity of plants there. Swamps contain woody plants and shrubs, including water-tolerant trees. Marshes have only grasses, cattails, and other herblike plants.

*Swamps and marshes are both wetlands, covered with water most of the time. The main difference between the two is the presence of flood-tolerant trees in swamps.*

The third type of wetland is bogs. What separates bogs from swamps and marshes is actually the presence of plants that lower the pH of the water, such as sphagnum moss and tamarack trees. The water in bogs is acidic, usually a pH of around 4.0. This is low enough to inhibit the growth of decomposers like fungi and bacteria, preventing dead plant matter from being broken down at the normal rate. Instead, the plant matter piles up at the bottom of the bog, forming a large, heavy mat. This partially decomposed plant matter is called peat. This is where the large cubes of peat moss in the hardware stores come from.

# Coastal Ecosystems

The ecosystems along the borders of oceans and continents have characteristics all their own. The interaction of the salty marine water and the land can produce a wide variety of conditions.

Some coasts are windy, with huge crashing waves; others are incredibly mild and calm. The wave action has a multitude of effects on the coast, creating soil that ranges from grainy sand to large boulders. Additionally, some coastal ecosystems also have a source of freshwater nearby, interacting with the marine water to produce an entirely unique concentration of dissolved solids.

Perhaps the biggest influence on coastal ecosystems is the tides. The ocean level will predictably rise and fall, depending on the gravitational pull of the moon at any given point in its orbit around the Earth. Every coastal ecosystem has an intertidal zone, named such because it's submerged during high tide, then dry and exposed during low tide.

## Marine Wetlands

Marine wetlands are classified in much the same way as the ones on land. Salt marshes have no trees and are covered with similar species of reeds and grasses. The main difference, of course, is the presence of saltwater instead of freshwater.

Mangrove swamps favor the growth of trees. Mangroves are trees that are able to grow in soil submerged under saltwater. These trees are unique in their ability to survive the high salt levels and tides, and are found only in these ecosystems.

When a coastal ecosystem is partially enclosed and has at least one river flowing into the sea, it is called an estuary. Estuaries are unique in that their water is neither fresh nor truly salty—it's brackish. As a result, you have an intermixing of organisms like no other aquatic ecosystem. The part of the estuary nearest the mouth of the river will be close to freshwater in its salt content. Meanwhile, the part closest to the actual ocean will have nearly the same salt content as marine water. The salt level in the middle varies.

Estuaries are one of the most biologically productive aquatic ecosystems known on Earth. Plankton and other photosynthetic organisms thrive here, nourished by the influx of sediment nutrients dissolved in the river water and the plentiful sunlight exposure.

## Beaches

Beaches are all characterized by one common feature: the presence of sand. The chemical composition of sand varies, but it's made of crystal-like shards of mineral. In fact, if you look at sand very close, it looks like tiny pieces of glass or rock. This hints at its formation, as all sand was once part of a larger rock or shell that has been eroded by waves into much smaller particles.

One of the biggest clues to the origin of sand is its color. White sand is made primarily of calcium carbonate. This mineral is found in the shells of living organisms. White sandy beaches are most often found near coral reefs, and are indeed the tiny remnants of coral that has been broken down by waves or the digestive systems of large, hungry fish. Brown sand is the eroded remnants of rocks found on older continents. Black sandy beaches are created when waves break down young volcanic rocks.

The life in sandy beaches, as in other coastal ecosystems, is very much influenced by the coming and going of the tides. Organisms that survive in the intertidal zone will often bury themselves in the sand to avoid predators and desiccation in the hot sun.

 **CASE STUDY**

From the human perspective, sand seems like a pretty homogenous mixture, where every particle of sand is identical. However, viewed up close, grains of sand are as unique as snowflakes.

Sand is the result of erosion, so the exact composition depends on exactly what was eroded to make it. Some grains of sand are fragments of crystals or minerals from the sea. Others are fragments of shells or coral, each bearing part of the design of the original structure.

## Mudflats and Rocky Shores

The same wave erosion that created the different types of beaches can create other types of coastal ecosystems as well, depending on the ferocity of the waves.

Mudflats, also known as tidal flats, are found in sheltered coastal areas where the wave action is very weak. Instead of depositing sand, the waves wash up much finer silt, mud, or partially decomposed plant and animal matter. This brew is perfect for the growth of many insects and larvae, which in turn attract great numbers of birds looking to feed on them.

On the opposite end of the spectrum, rocky shores are found in areas with strong wave action. The shore is littered with a variety of large rocks. This might seem the least hospitable to life, but rocks are the preferred surface for many filter-feeding shellfish. Mussels and clams will secrete proteins that glue them tight to rocks within the intertidal zone. They filter out plankton and small particles of dead matter when high tide is in, then withdraw into their shells during low tide.

# Marine Ecosystems

As mentioned at the beginning of this chapter, the vast majority of Earth's water is found in the oceans. Yet, in spite of all that volume, the most productive and diverse ecosystems are considered to be the coastal ones.

Much of the ocean is actually pretty devoid of life. Remember, the basis of most food chains is the producers: photosynthetic bacteria, plants, and algae. Move away from the nutrient-rich sediment runoff from the continents, and these producers do not flourish in nearly the same numbers.

## The Open Ocean

Paradoxically, the open ocean is very desertlike, at least with the amount and diversity of life it supports. The limiting factor in deserts is precipitation, but in the open oceans, it is nutrients for the plankton.

The relative lack of life in this ecosystem is literally crystal clear when you observe the water. Aquatic ecosystems that are rich in life can be difficult to see through. Thick mats of algae and plankton can completely obscure the bottom layers. However, in many areas of the open ocean, you can see as far down as light can penetrate. The populations of photosynthetic plankton are simply too sparse to be visually apparent.

The open ocean certainly is not devoid of life. Large schools of fish, whales, and other aquatic organisms pass through this ecosystem regularly. However, much like the arid deserts of the continents, these organisms are constantly on the move in search of food.

## Ocean Deep

The entire character and diversity of aquatic ecosystems change drastically as you travel down toward the bottom of the ocean. The water pressure increases so much that humans can't safely travel that deep without an armored submersible. The water temperature drops as energy from the sun becomes more and more distant.

Most importantly, the water becomes dark. The upper layer of the open ocean, called the euphotic zone, supports photosynthesis. This is where the most productive ecosystems exist, including the coral reefs and estuaries. The middle layer, called the disphotic or twilight zone, still has some light, although it is very faint. Only a few of the colors present in white light can actually penetrate down this far, red being one example. Finally, the bottom layer, or aphotic zone, is completely devoid of light.

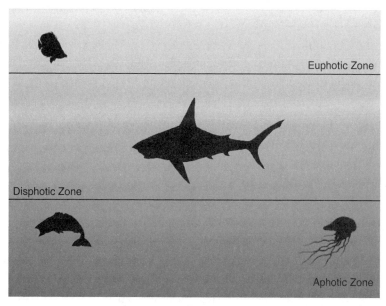

*The biggest transition between these three zones is light. Photosynthetic organisms concentrate in the euphotic zone and decomposers in the aphotic zone, while predators tend to move between the zones.*

There can be no photosynthesis at the bottom of the ocean. Yet life still exists on the bottom, supported instead by a constant trickle of energy and nutrients from dead matter and waste falling from above. This is called "marine snow" because it looks like little white flecks falling from above.

This marine snow is the basis of the entire ecological pyramid at the bottom of the ocean. Producers are replaced by decomposers and scavengers, which feed off these particles of decay. Primary and secondary consumers will in turn feed off the decomposers and scavengers.

To add further diversity to the ocean bottom, there are a few places where an entirely new type of food chain will spring up, fed not by sunlight or decaying matter but by heat from the Earth's mantle. Hydrothermal vents are fissures in the ocean floor from which heated water escapes. Some of the fissures take the shape of chimneys, emitting compounds that resemble white or black smoke. The producers in this ecosystem are able to create their own food from the heat energy released by the fissure. A community of organisms will spring up around them, completely dependent on the chemicals present in the super-heated water. If the fissure were to stop flowing, the entire community would perish.

## Coral Reefs

Coral reefs are oases of life in the desert of the open ocean. They are found in latitudes close to the equator, so the water is warm and sunlight is plentiful. Essentially, these could be ideal conditions for a thriving community, assuming enough producers are present.

The primary producers in this ecosystem are actually an example of mutualistic symbiosis—two organisms that live close together, benefiting each other. Coral contains animals called polyps and photosynthetic algae. The polyps produce a protective calcium carbonate skeleton. The algae have the ability to produce their own food from sunlight. Together you have a formidable pair and the keystone species of this entire ecosystem.

 **CASE STUDY**

The mutualistic relationship between coral polyps and the algae in coral reefs is a very delicate one. Under certain conditions—increase in water temperature, decrease in water pH, changes in salinity, and so on—the reef loses its ability to provide needed nutrients for the algae to conduct photosynthesis. The algae are ejected from the coral, and the polyps, deprived of their primary source of food, die. All that's left is the white calcium carbonate skeleton, leading to the "bleached" appearance. If the current rate of coral reef loss continues, it's estimated that 70 percent of all reefs worldwide will be gone in the next 30 to 40 years.

Every species that resides in the reef owes part of its survival to the coral. Some fish feed directly on it, others live within the maze produced by the outgrowth of the polyps. There are plenty of secondary and tertiary consumers as well, from octopi to sharks. Altogether, coral reefs have biodiversity rivaling that of rainforests.

While we tend to think of all waters as being roughly identical, there is a tremendous amount of disparity between them. Differences in temperature, light, and salt levels can all influence what species can and cannot survive there.

## The Least You Need to Know

- Aquatic ecosystems are primarily differentiated by the presence of salt, temperature, and nutrient availability.
- Nutrient levels in water tend to be higher near land due to soil runoff from rivers and rainfall.
- Swamps are wetlands with trees, marshes have no trees, and bogs have acidic water that inhibits bacterial decomposition.

- The different substrates found in shoreline ecosystems—sand, rock, and silt—are primarily the result of greater or less wave action.

- Coral reefs and estuaries are considered the most diverse of all the aquatic ecosystems, while the open ocean is much less productive.

- The bottom of the ocean is devoid of sunlight, so the food webs are centered around other sources of energy, such as detritus from above and hydrothermal vents.

# The Movement of Energy and Matter

Life has existed on Earth for billions of years. Millions of different species inhabit thousands of ecosystems all across the biosphere. All this life, in all its diversity, does have two very basic things in common: it's all made of matter, and it all uses energy.

Matter is everything around you—anything made of atoms and molecules. By definition, matter is anything that has mass and takes up space. Everything on Earth, living or not, is made of matter. The amount of matter on Earth has changed very little over the 4.5 billion years it has existed as a planet.

Energy is the ability of some physical system to do work. It's expressed in many different ways: light, heat, movement, and so on. Energy is vitally important for life to exist, and the energy on Earth, unlike matter, isn't a constant quantity.

In this chapter, you'll learn about the basic principles that guide the flow and movement of these two basic forces— matter and energy. Each is essential for life to exist, but is utilized and passed along in very different ways.

## In This Chapter

* The flow of energy through an ecosystem
* Different sources of energy used in living organisms
* A look at the laws of thermodynamics
* The cycles that allow matter to be used and reused
* The all-encompassing major elements of life

# Energy Flows

Energy can exist in two different states. Energy that's stored is *potential energy*. This type of energy exists as a result of a difference in position. For example, water at the height of the reservoir in a hydroelectric dam has a great deal of potential energy. By the time it has worked its way through the turbines of the dam and exited the other side, it has much less potential energy.

Energy can also exist in a form with movement and force. This is *kinetic energy*. A spinning turbine, a moving car, and an explosion are all different examples of kinetic energy.

 **DEFINITION**

**Potential energy** is stored energy that exists in certain molecules or as a result of the position of an object. Examples include sugars, fats, proteins, and burnable fuel such as gasoline. **Kinetic energy** is energy an object possesses due to its motion. Examples include single-celled organisms swimming in a pond or a steam-powered locomotive.

## Forms of Energy

Energy comes in six forms: chemical, electrical, radiant, mechanical, nuclear, and thermal. The exact definition of these forms isn't important. What is more relevant is the fact that energy can be converted between any two forms.

There are two important laws dealing with the movement of energy, referred to as the laws of thermodynamics. The first law states that energy is always conserved. Whenever a chemical reaction or energy conversion occurs, the energy input is equivalent to the energy output. The second law states that after a conversion or reaction, the potential energy will be reduced.

Don't these two laws contradict each other? Consider a car running on gasoline. The gasoline represents the chemical (potential) energy that's put into the system of the car. As the car runs, this chemical energy is converted into the actual movement of the car. Eventually, the car runs out of gas. What happened to the energy? Place your hand on the hood after a long drive. What do you feel? The answer is heat.

While the amount of energy didn't change in this conversion, the quality of energy did. The energy will be gradually released as heat—by the engine, the brakes, or the friction of the wheels against the road. There's no practical way to recapture all this energy; it will dissipate into the atmosphere and eventually into space. Unless you fill up the gas tank with more chemical energy, the car will no longer move on its own.

A British scientist by the name of C.P. Snow summarized these first two laws simply:

1. "You cannot win." In other words, you can't produce energy from nothing; it's conserved.

2. "You cannot break even." There's no way to return to the original energy state after a reaction.

There's actually a third law, which he summarizes as "You cannot get out of the game." This is in reference to a concept called absolute zero, which is the complete absence of energy. Absolute zero is unattainable. Energy is everywhere in the universe; even the weakest bit of light from a distant star is still energy.

## Energy in Organisms

Just like the engine of a car, every process inside living organisms, down to the cellular level, requires energy. Instead of gasoline, the potential energy is represented by complex organic molecules of sugar, proteins, or fats. As the organism lives, and as its cells do their work, that potential energy is converted into kinetic energy and then eventually is lost as heat.

Potential energy is stored in living organisms as chemical energy. Specifically, it's stored within the bonds between atoms in large molecules. This provides a way for the organism to access the energy later. When its body cells break down those molecules, the energy is released and can be used.

Organisms generally store energy in three forms: as complex carbohydrates, as protein molecules, or as fats. Complex carbohydrates include molecules like starch. Think about examples of foods high in starch: potatoes, pasta, and bread—all are plant-based foods, and they all have a lot of energy (calories). Carbohydrates are named as such because they're made of only three elements: carbon, oxygen, and hydrogen.

$$
\begin{array}{l}
\text{H} \quad\ \, \text{O} \\
\ \ \diagdown \text{C} \diagup\!\!\!\!= \\
\ \ \ \ \ | \\
\text{H-C-OH} \\
\ \ \ \ \ | \\
\text{HO-C-H} \\
\ \ \ \ \ | \\
\text{H-C-OH} \\
\ \ \ \ \ | \\
\text{H-C-OH} \\
\ \ \ \ \ | \\
\text{CH}_2\text{OH}
\end{array}
$$

*Glucose, a simple carbohydrate often referred to as blood sugar, is a small organic molecule made from carbon, hydrogen, and oxygen.*

Fats and oils are found in both animals and plants. Canola oil, olive oil, and lard are all different examples of energy storage in molecules of fat. Storing energy as fat has two advantages over using carbohydrates. First, fat acts as an insulator. This is a big advantage for animals living in Arctic or Antarctic ecosystems. Second, fat is more energy dense than carbohydrates; it can store more than twice the amount of energy per gram.

*Lipids are made of three fatty acids joined together. Saturated fatty acids, like the one at the top, are straight and form solids. Unsaturated fatty acids bend at their double bonds and form liquids.*

Proteins are the last type of organic molecule that relates to energy. Organisms usually don't make proteins specifically to store energy; they're involved in lots of other important body functions, like movement, protection, support, and cell-to-cell communication. However, when one organism consumes another, these proteins can be digested and converted into chemical energy. Proteins are most associated with animal sources of food, but there are plant sources as well, such as legumes.

*Proteins are large molecules made of combinations of different amino acids, such as alanine, shown here.*

## Energy in Ecosystems

Think about how food chains and food webs are drawn. The arrows point from the prey organism toward the predator or parasite organism. This shows the flow of energy through the ecosystem. Does this follow the laws of thermodynamics? The first law is followed because no energy is produced; rather, it's converted from the chemical energy in one organism (the prey) into kinetic or chemical energy in another (the predator). The second law is also followed, because on average, only about 10 percent of the energy is actually converted from one level to the next.

Given this tremendous amount of inefficiency, the 90 percent of chemical energy that's wasted as heat with each energy conversion, how are ecosystems sustainable? What's the continual source of energy that keeps this system going? The answer for most ecosystems is the sun. The sun provides the constant flow of light energy needed to fuel photosynthesis, forming the basis for the entire *producer* trophic level, which in turn supports all the other *consumer* trophic levels.

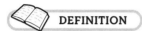 **DEFINITION**

> A **producer** is any organism that can produce its own molecules of high-potential energy from a source like sunlight or heat. A **consumer** can't produce its own food directly and must ingest or absorb other living tissue.

Are there any communities independent of the sun? If you recall from our tour through the aquatic ecosystems in Chapter 7, communities living at the bottom of the ocean have no contact with sunlight and rely instead on detritus that falls from above. However, without the sun, the communities in the upper levels of the ocean would not exist, nor would their detritus.

The only truly independent communities are those living off the hydrothermal vents in the ocean floor. The organisms there extract all the energy they need from the heat and chemicals these vents release. If the sun were to suddenly extinguish, these life forms would probably be the only ones on Earth able to survive.

# The Carbon Cycle

Visualize a spaceship traveling through space. The spaceship has no ability to mine or extract any minerals or resources from space. Whatever matter and resources exist on the spaceship are all the inhabitants have to survive on. Whatever waste the inhabitants of the spaceship produce must be recycled or reused somehow. Sure, they could just eject all their garbage into space, but wouldn't that produce shortages of material?

 **A LOOK BACK**

> Spaceship Earth is actually an environmental worldview first ascribed by an 1879 book titled *Progress and Poverty*. The relevant passage states, "It is a well-provisioned ship, this on which we sail through space. If the bread and beef above decks seem to grow scarce, we but open a hatch and there is a new supply, of which before we never dreamed. And very great command over the services of others comes to those who as the hatches are opened are permitted to say, 'This is mine!'" The basis of this worldview is that the Earth has a limited amount of resources, and it's in the best interest of everyone living here to use them responsibly so the society may continue to survive.

In many ways, the Earth is like this spaceship example. Very little matter enters and leaves the Earth. Sure, there have been some massive meteor strikes, but these are pretty rare. All the matter on Earth—the minerals, fossil fuels, elements, and compounds—is basically there in a fixed, finite amount. Likewise, any waste we produce remains. The only thing that enters and leaves the Earth in large amounts is heat.

The point is, while energy can flow one way through an ecosystem, it's not sustainable for matter to do so. Matter must be continually reused, over and over again. Life on Earth has been able to exist for billions of years because it recycles.

# The Importance of Carbon

The number of elements essential for life is actually pretty small. The most prevalent ones are given the nickname "CHNOPS": **C**arbon, **H**ydrogen, **N**itrogen, **O**xygen, **P**hosphorus, and **S**ulfur. There are others, called trace elements, but these exist in much smaller amounts.

For life to be sustainable on Earth, there must be a way for each of the elements to be used and then returned to the soil or atmosphere to be used again. These are called biogeochemical cycles because the matter can go between living matter (bio) and the Earth's crust (geo), taking many different chemical forms along the way.

Carbon is often called the central element of life. Carbon has a couple of unique properties. First, it can form four separate chemical bonds with other elements at the same time. It can also bond with itself, creating long chains of molecules (e.g., proteins, carbohydrates, and lipids). Carbon can also form double and triple chemical bonds. Long story short, carbon is an incredibly versatile atom that can build countless different types of molecules.

Where is carbon found on Earth?

## Sources of Carbon on Earth

| Carbon Pool | Estimated Amount (Gigatons) |
| --- | --- |
| Fossil fuels | 4,000 |
| Oceans | 38,000–40,000 |
| Soil organic matter | 1,500–1,600 |
| The atmosphere | 750 |
| Plant matter | 560 |

*Source: Global Carbon Reservoir Data, esd.ornl.gov/projects/qen/carbon2.html*

## Photosynthesis and Cell Respiration

Most of the carbon on the Earth that's accessible to living things at any given moment exists as carbon dioxide in the atmosphere and the oceans. This carbon dioxide becomes incorporated into complex organic molecules of living things through the process of photosynthesis.

Photosynthesis uses energy from the sun in the form of visible light to convert carbon dioxide, which consists of very low-energy molecules, into glucose, a high-energy carbohydrate. An additional byproduct of this reaction is oxygen. This gas, exhaled by the plants as waste product, is actually an essential ingredient for cell respiration, the cellular process animals use to metabolize carbon.

Cell respiration looks kind of like a mirror image of photosynthesis. While the two processes aren't exact opposites of each other, the end products essentially are. Cell respiration takes in glucose and breaks its chemical bonds, releasing energy that cells can use. The leftover carbon, now carbon dioxide gas, is exhaled as a waste product to be used later by another plant, and the cycle continues.

## Decomposition and Combustion

Photosynthesis accounts for the growth of plants. As they absorb more sunlight and carbon dioxide, they produce more sugars, which then can be used to build the different structures of the plant: leaves, stems, roots, and so on. Cell respiration accounts for animal digestion—breaking down sugars from plants, extracting the chemical energy, and releasing the carbon back into the air. This isn't the only means by which carbon can be released, however, as not every plant is eaten by an animal.

Many plants simply die. The causes vary, from drying out to a fungal infection to injury caused by a lightning strike. Nevertheless, a dead plant, rich in carbon sugars, now rests on the ground. This is where decomposers come into play. Bacteria, fungi, and other microscopic organisms will begin their own form of consumption and digestion, breaking down the plant until its constituent elements have been returned to the earth. The actual waste products may be solids that enter the soil, or gaseous carbon dioxide or methane that enters the air.

 **CASE STUDY**

The carboniferous ("coal-bearing") period was a geologic time period about 360 million years ago. Much of the limestone (calcium carbonate) and coal in the Earth's crust is believed to have originated during this time. How? Plants had evolved the ability to produce lignin, the stiff, carbon-based fiber that forms their bark and trunks. No organisms had yet evolved the ability to digest this substance, so it literally accumulated in massive heaps on the ground. Some of these materials were buried, eventually being transformed by millions of years of heat and pressure into the fossil fuels we utilize today.

In dry ecosystems, the decomposers may never really have a chance to perform their work. The overall lack of moisture inhibits their growth and increases the chance of a different kind of chemical reaction: combustion. When a fire occurs, whether in a forest or grassland, the molecules within the plants are consumed and converted into new forms: carbon dioxide and ash. Again, once the fires have burned out, the necessary elements of life have been returned to the atmosphere or soil to be used again.

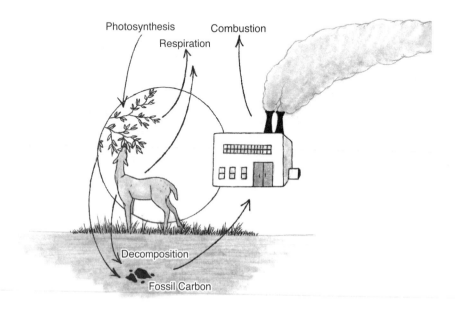

*Carbon continuously moves back and forth between the atmosphere, living organisms, and the Earth's crust. It may be part of living organisms, fossil fuels, or carbon dioxide gas.*

# Other Biogeochemical Cycles

While carbon is certainly an important element to living organisms, it's far from the only one that must be recycled. The remaining CHNOPS elements also have their own cycles, ensuring their availability for generation upon generation of living organisms.

The movement of water actually accounts for two of the CHNOPS elements: hydrogen and oxygen. Water is the primary source of hydrogen in living organisms. It has its own cycle, which we will cover in Chapter 21.

Oxygen, in addition to being part of the water cycle, also moves between organisms through other cellular processes. The same is true for nitrogen, phosphorus, and sulfur.

# The Oxygen Cycle

By mass, oxygen is actually the single most prevalent element in the human body. This is mostly due to water, but elemental oxygen is also an important necessity for life.

Oxygen is a diatomic gas element, meaning it bonds to itself. It's colorless and odorless. During the early days of the Earth, the atmosphere actually had very little oxygen. This changed when a group of organisms called cyanobacteria began to use sunlight to convert carbon dioxide and water into simple sugars. Oxygen was produced as a waste product and began accumulating in the atmosphere.

Many other organisms have evolved since the original cyanobacteria, including all the land plants we now know. They follow the same basic formula:

$$CO_2 + H_2O + sunlight\ (energy) \rightarrow C_6H_{12}O_6\ (glucose) + O_2$$

Of course, this is only half the picture. The oxygen that's given off as a result of photosynthesis is used to metabolize these same sugars in animal cells. Does this sound familiar? It should, because this interplay is very much the same as with the animal and plant parts of the carbon cycle.

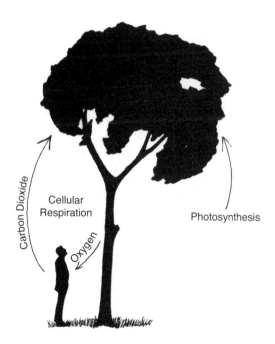

*The two main processes that move oxygen throughout the Earth are cellular respiration (which takes place in all living cells) and photosynthesis (which takes place in producers only).*

# Nitrogen, Phosphorus, and Sulfur

In fact, the rest of the nutrient cycles are actually pretty similar to the carbon cycle. Phosphorus, for example, is an important component in many organic compounds in both animals and plants. ATP, the basic form of energy used by all cells, incorporates three phosphorus atoms.

When animals and plants die or produce waste, *detritivores* and *decomposers* metabolize the remains, releasing phosphorus back into the soil. At this point, it may run off into water, become incorporated into deep layers of soil and sediment, or be reabsorbed by plant roots.

 **DEFINITION**

> A **detritivore** is an organism that's able to metabolize nonliving fragments of organic matter, such as animal waste. A **decomposer** is an organism that's able to break down dead or decaying organisms into smaller molecules.

An important difference between the phosphorus and carbon cycles is that phosphorus never enters the atmosphere. The biggest sink of nitrogen, on the other hand, is the atmosphere. In fact, nitrogen by far makes up the biggest part of the air we breathe—more than 75 percent. This nitrogen can be integrated into living things through bacteria, which have actually formed symbiotic relationships in the roots of certain plants. Legumes, such as soybeans and clovers, are one.

Nitrogen and phosphorus are two of the essential nutrients plants need to reach their maximum growth potential. If you look at the key ingredients on any commercial plant food or fertilizer you will see these two elements, along with potassium. Each of these elements is reincorporated into soil when plants die and decompose. However, many homeowners dispose of their leaves and grass clippings instead of recycling them, and then use synthetic fertilizer to grow a healthy lawn.

The last major element of life, sulfur, can exist both in the atmosphere and as part of the soil. Sulfur enters the atmosphere when some fuels (coal, especially) are burned. It's also a major component of the gases released by volcanic eruptions. This sulfur combines with water and can re-enter the soil, where it's metabolized by bacteria and absorbed by plants. It then becomes an important component of many plant and animal protein molecules.

The important idea to remember from this chapter is the contrast between how the environment handles energy and matter. Energy, by its nature, cannot be continuously reused. There is no 100 percent efficient system. Thus, a continuous flow of new energy (sunlight) is needed to sustain life. All forms of matter can be reused without degradation. The amount of matter on Earth has changed very little since life began.

## The Least You Need to Know

- The Earth is an open system for energy; most of the input comes from the sun, while the output is heat loss.
- Energy cannot be created from nothing; a food web must have a constant input, such as the sun or chemicals from the Earth's mantle.
- Energy flows through an ecosystem, with some lost as heat during each transformation. Matter is reused and recycled without any such loss.
- Each of the elements of life (CHNOPS—carbon, hydrogen, nitrogen, oxygen, phosphorus, and sulfur) has a set of living and nonliving mechanisms to move and cycle them throughout the Earth.

# Meeting Human Needs

Part 3 begins a gradual introduction of human influences on all the ecosystems, species, and resources throughout the Earth. We begin by examining human population growth, projections of our growth, and how this growth varies from country to country. The focus then moves to the influences that economics has on the resources we use, the value we place on the environment, and the decisions we make in how to use it.

Next we look at providing for the needs of human nutrition. Growing crops and raising animals have changed a great deal in the last 50 years, becoming more and more mechanized. The modern process of cultivating and harvesting plant-based and animal-based foods is outlined, along with some of the environmental hazards inherent in the production process.

Finally, I trace the different sources of our seafood, including wild-caught fish taken by poles or trawl nets and massive farming operations along the coast. A big theme of this chapter is the consequences of harvesting these animals at too fast a rate.

# The Human Population

We've spent the last few chapters covering all the patterns of growth and laws of energy and matter that govern how ecosystems develop. These rules and patterns also apply to the human population.

We now know that the amount of matter on Earth essentially remains unchanged over the course of millennia, and any matter-based resources we need to survive are finite. Natural communities of organisms adapt to this by cycling each of the major elements of life through themselves, the atmosphere, and the soil. Yet humans produce tremendous quantities of garbage that is locked in underground landfills and removed from these normal cycles.

We also recognize that populations usually grow in a sigmoidal (S-shaped) pattern. The growth rate slows as the population encounters limiting factors like disease, predation, and lack of resources. Yet the human population has surpassed 7 billion, by far the most prolific of any other vertebrate on Earth.

The growth and proliferation of our species seem to defy nature. Any other species would have crossed the carrying capacity, experienced some sort of population crash, and suffered the consequences of a depleted ecosystem. Somehow, our population has largely avoided this outcome.

## In This Chapter

- The changing patterns of human population growth
- Factors that affect the population dynamics of a country
- How population pyramid graphs are made
- The stages of the demographic transition model

# How Did We Get to Seven Billion?

To understand the dynamics of our own population, a quick history lesson is in order. We haven't always grown exponentially, at the rate we are now. We haven't always been able to evade the density-dependent limiting factors to growth.

Recall from the very first chapter that the earliest humans were hunter-gatherers. Our species was no different than any other animal. We travelled in groups, following herds of large herbivores that we hunted, and gathering plant material when it was available. Any change that affected the lower levels of the food chain, producers or primary consumers, affected our species as well.

Over time, as we developed more and more advanced technologies in the fields of medicine, engineering, and agriculture, the entire dynamics of our population changed. We began to exploit nature to benefit our species, instead of being beholden to it. This accumulation of changes has led us to the extraordinary population size we now maintain.

## Before the Industrial Age

Before we talk about the changes in growth of these early human populations, a little disclaimer is in order. Obviously, there were no massive census databases keeping track of the human population. There were no people with clipboards going door to door tracking changes in the numbers, age range, and ethnicities of the people.

All world population data prior to 1950 exists as historical estimates. These estimates are proposed population sizes based on historical records and archeological discoveries.

Here is a set of data based on the lower-level estimates, provided by the U.S. Census Bureau.

### Historical Estimates of World Population, 10000 B.C.E. to 1800 C.E.

| Year | Population Estimate |
| --- | --- |
| 10000 B.C.E. | 1 million |
| 4000 B.C.E. | 7 million |
| 1000 B.C.E. | 50 million |
| 1 C.E. | 170 million |
| 500 C.E. | 190 million |
| 1000 C.E | 254 million |
| 1500 C.E. | 425 million |
| 1800 C.E. | 813 million |

The human population certainly grew during this time, to an impressive 813 million worldwide. This growth, as seen on the following graph, was relatively steady, although the human race did encounter a few speedbumps as its size increased.

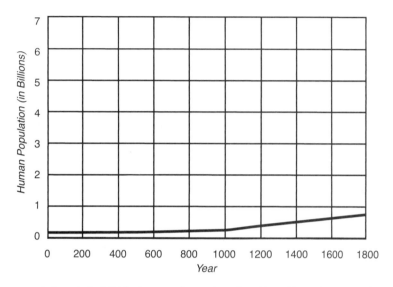

*Human growth through the Middle Ages was slow and steady, taking a linear shape. Even large-scale famines and diseases had relatively little effect on the overall growth rate.*

## Bumps in the Road

From the Agricultural Revolution through the Middle Ages, humans did a remarkable job of spreading throughout the biosphere. Different tribes and populations were found on every continent except Antarctica. They exercised some control over their environment by constructing permanent shelters and cultivating animal and plant crops. However, they were not fully exempt from the limiting influences of the environment.

Remember some of the density-dependent limiting factors we covered earlier. Which apply most to humans? While there are a few animals that will prey on humans, it is pretty rare. Disease, however, was very much a factor during this time.

The Black Death was a disease caused by a bacterium. Symptoms of the disease vary, but include buboes, which are swelling of the lymph glands, fever, vomiting, and eventually, death. There was no effective treatment, and the death toll was enormous. It's estimated that in the fourteenth century, between 75 and 200 million people died. This included about half of Europe's total population!

> **CASE STUDY**
>
> The plague is far from the only epidemic disease that has impacted the human population. Smallpox is a virus that causes a rash, then fluid-filled blisters all over the skin of an infected victim. The disease was especially deadly to populations that had never been exposed to it. The death rate was around 80 percent in Native American people during the eighteenth century. This disease alone had little significant impact on human population growth, but combined with the other epidemics of the time did slow the growth rate considerably.

In spite of the incredibly widespread nature of this disease, it slowed but did not stop the overall growth rate of the human population, and the first billion was reached around the time of the Industrial Revolution.

## Population Explosion

In all, it took tens of thousands of years to reach humankind's first billion. Growth was slow and steady, taking more of a linear shape when graphed.

During the nineteenth and twentieth centuries, a multitude of technological advances changed the entire dynamic of our population growth. Electricity, running water, sewage treatment, vaccines, and antibiotics combined to greatly improve our life expectancy.

The expected age of death for any given individual prior to the Industrial Revolution was pretty consistent. The single biggest factor that influences life expectancy is actually infant mortality rates. Imagine if you have a thriving village where adults surviving into their fifties are not unusual. However, due to the lack of modern medicine, many children die before the age of five. Averaged together, this makes for a life expectancy of somewhere around 30.

### Life Expectancy Through the Industrial Revolution

| Time Period | Example Year | Life Expectancy |
| --- | --- | --- |
| Bronze Age | 1000 B.C.E. | 26 years |
| Ancient Greece | 400 B.C.E. | 28 years |
| Medieval Britain | 1200 C.E. | 30 years |
| Pre-Industrial Britain | 1700 C.E. | 25–40 years |

With the advent of vaccines and antibiotics, the highly fatal childhood diseases became much less so, and global life expectancy reached 67 by the year 2010.

## Slowing Down Again

As infant mortality fell and life expectancy rose by the mid-twentieth century, the human population began to grow at an increasingly faster rate. For example, a medieval couple giving birth to five children might see only three of them survive into adulthood. Within a few centuries this changed, and all five children were likely to survive. This was great news and reflected a tremendous improvement in human health during this time period.

However, old habits die hard. Industrial age couples certainly did not stop having children, nor did they attempt to have fewer. As a result, the rate of human population growth increased dramatically.

Let's look back at our historical population estimates from the census data available.

### Historical Estimates of World Population, 1800 c.e.–2000 c.e.

| Year (c.e.) | Population Estimate |
| --- | --- |
| 1800 | 813 million |
| 1850 | 1.13 billion |
| 1900 | 1.55 billion |
| 1910 | 1.75 billion |
| 1920 | 1.86 billion |
| 1930 | 2.07 billion |
| 1940 | 2.40 billion |
| 1950 | 2.56 billion |
| 1960 | 3.04 billion |
| 1970 | 3.71 billion |
| 1980 | 4.45 billion |
| 1990 | 5.29 billion |
| 2000 | 6.01 billion |
| 2010 | 6.86 billion |

Within 200 years, the human population increased by more than eight times. The first billion took over 70,000 years, but the second only took 130. A third billion was added after 30 years, a fourth in about 25. While the population has always been increasing, since the Industrial Revolution its *rate of growth* is also increasing. Look at the graph:

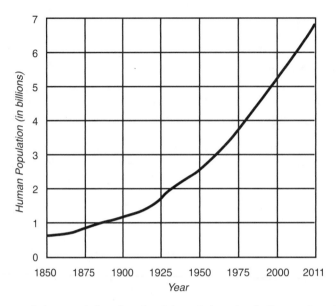

*Human population growth since the Industrial Revolution takes the J-curve shape. This is exponential growth, where the actual rate of growth (the slope of this graph) is also increasing.*

The human race is experiencing exponential growth, the kind of growth that, in other species, invariably leads to an overshoot of the ecosystem's carrying capacity and a dieback. Are we overdue for a population crash, or do we have the capacity to rein in our own growth before it is too late?

## How We Measure Growth

The sheer size of our population is staggering and difficult to conceptualize. Instead of looking at total population, let's just focus on the actual growth rate.

Data for annual growth rates prior to 1950 is difficult to come across, as world census data is incomplete. However, starting in 1950, we can get a pretty good idea of how this growth rate has been changing. We can also use this data to calculate doubling time, an estimate of how long it would take our population to double at that rate. The equation for doubling time is:

Doubling Time = 70 ÷ % Growth Rate

So if a population is growing at 7 percent per year, it has a doubling time of 70 ÷ 7 = 10 years. Fortunately, our growth rates aren't quite that fast.

## Annual Growth Rate and Doubling Time, 1950–2010

| Year | Growth Rate | Doubling Time |
|------|-------------|---------------|
| 1960 | 1.35 | 52 years |
| 1970 | 2.09 | 34 years |
| 1980 | 1.82 | 38 years |
| 1990 | 1.60 | 44 years |
| 2000 | 1.26 | 56 years |
| 2010 | 1.12 | 63 years |

This data tells a much different story. Since 1970, the growth rate has actually declined, and the doubling time has increased. Our rate of growth is actually starting to slow down. Why is this happening, and is this true for every country around the world?

# Demographics: Looking Inside the Numbers

All the population trends we've looked at so far have been worldwide. While this is helpful in giving a big-picture view of what is happening with the human population, there is actually a lot of variation from country to country.

Currently, there are countries in the midst of a population explosion. If placed on a graph, their growth would appear to be very much exponential. Other countries have stable populations experiencing very little change, while still others are actually shrinking in size.

The study of *demographics* looks at these individual differences—gender, age distribution, economics, and other variables that influence a society and its growth rate.

As an example, let's look at the influence that economics has on *life expectancy*. Now, you would expect that wealthier countries on average would have a lower *infant mortality rate* and greater life expectancy. Indeed, that is the case, but only up to a point.

When household income rises to about $5,000 per year, life expectancy increases dramatically. However, there is a ceiling to this benefit, as most wealthy countries have been unable to break the 80-year life expectancy benchmark. There are certainly a lot of other variables that come into play, not to mention our own natural life span limits.

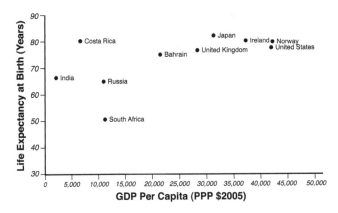

*Life expectancy is only somewhat correlated with income. Countries with a GDP per capita below $5,000 per year usually (though not always) have lower life expectancies.*

## Kids and Family Size

Population might seem like a relatively simple thing to measure. The equation for calculating population change is no different than for any other organism:

Change in Population = (Births + Immigration) − (Deaths + Emigration)

However, with humans, reproducing and having families is not strictly an instinctual, biological function. There are a lot of psychological and sociological factors that influence the number of children a person has. As a result, there are a few other measurements we can use to give a clearer picture of what is happening within a country.

*Fertility rate* is the number of births per thousand women per year. Most demographic measurements are actually per thousand, just to keep the numbers smaller and more manageable. In 2010, the fertility rate in the United States was 63.4. *Total fertility rate* takes this number and distills it down to the average number of births any given woman will have in a country. Again, using the United States as an example, our total fertility rate is 2.1. Remember, this is an average, so some women will have fewer children and others will have more.

 **DEFINITION**

**Demographics** is a study of all the characteristics of a human population. Important variables include **fertility rate,** the birth rate per thousand women, and **total fertility rate,** the average number of births per woman. Another variable is **life expectancy,** the average age a newborn will reach. This is most influenced by **infant mortality rate,** the number of infant deaths per thousand births.

# The All-in-One Population Graph

Two of the most important factors that demographers look at when studying a country are age and gender distribution. The relative population size of different age groups can give you an idea of how long the average person lives, how many adults are within reproductive age, and how many children they're having. Any differences in gender can give some insight into other sociological factors we'll discuss later.

Population pyramids are basically sideways bar graphs. Each 5-year age group has two sideways bars, one for male population and one for female population. Look at the following example, Ethiopia in the year 2012.

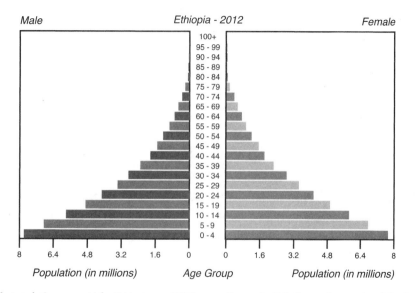

*The population pyramid for Ethiopia as of 2012, according to the U.S. Census international database.*

The pyramid graph might not seem to shed much light on a country's demographics at first, but consider the following questions whenever you look at one:

- What is the overall shape of the graph? Does it look like a pyramid, or is it more rectangular?

- Is there a dominant age group (children, adults, elderly) in the population?

- Compare the 0–4 age group with the top age groups. Do a lot of people seem to survive to these upper ages?

- Compare the male and female sides. They should be roughly equal. Are they?

Ethiopia's graph is definitely pyramid shaped, with the younger age groups being the dominant ones in the population. It would be reasonable to conclude that this country has a pretty high birth rate (accounting for the large base), a high fertility rate (over 4 children per woman), and a lower life expectancy (58) compared to other countries. Population pyramids can look wildly different from country to country, depending on the economic and demographic situation.

## Growth Rates in Developing Countries

As you may remember, developing countries are the least wealthy and tend to have the most acute environmental problems. For many of the people poverty creates a situation in which their short-term survival is at risk, and they make a lot of decisions that cause long-term damage in order to sustain themselves in the present day. Deforestation (from selling timber and clearing farmland), extinction (a result of poaching and selling exotic animal products), and soil erosion (caused by overfarming) are all commonplace in these countries.

Given the harsh living conditions, lack of opportunity, and severely depleted resources, one might expect the birth rate of these countries to be low, as parents anticipate an inability to care for many children. However, the opposite is true.

Developing countries demonstrate what is called a population paradox: they have the highest birth rates in the world despite not having the resources to care for their own people.

Unraveling this paradox is no simple matter. There is a series of variables called pronatalist forces that increase a person's likelihood of having children. Some of these are universal—for example, the pride and pleasure associated with having a family. However, many of these forces are most apparent in developing countries. For example, some people may have additional children to provide a means of support in their later years. Children are often expected to care for their parents when they get older or to take over the family farm.

There are also a lot of cultural roots behind pronatalist forces. There may be religious objections to the use of birth control. If the society is a male-dominated one, the choice to have more children may be completely out of the hands of the wife and instead lie with the husband and his family.

Many cultures also have a strong preference for sons. Men may have more rights and opportunities compared to women. Girls may be seen as a liability, such as when a dowry from the girl's family is given when a man agrees to take her as his wife. As a result, many husbands will pressure their wives to have many children to maximize the odds of producing male heirs.

To see all these forces in action, take a look at the population pyramid of India:

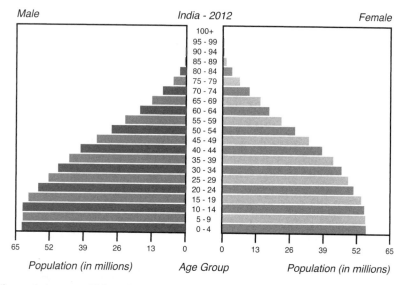

*The population pyramid for India as of 2012, according to the U.S. Census international database.*

What patterns do you see in this graph? Clearly, there is a much wider base of children compared to the middle-aged and elderly. There is also a noticeable difference between population size of males and females. There are disproportionately fewer females in the younger age groups. How can this be? The answer is sex-selective abortion. Some couples will have an ultrasound to determine the gender of their child, aborting (often illegally) girls and keeping boys.

### CASE STUDY

*Humraahi* was an Indian soap opera broadcast between 1992 and 1993. The central storyline of the soap opera was an adolescent girl who was forced into marriage and early pregnancy instead of pursuing her dream to become an attorney. Halfway through the series, the main character died giving childbirth. The remainder of the series focused on the other characters coming to realize the need to delay the age of marriage for girls and provide more opportunities for education. This soap opera, combined with other social outreach programs, was part of an ongoing effort to hasten the cultural shift in India toward more gender equity.

For developing countries the end result of all this is high fertility rates, growing populations, and increasing pressure on the already strained natural resources of the country.

## Growth Rates in Developed Countries

The demographic picture in developed countries is the polar opposite: fertility rates are much lower. Again, this is part of the population paradox, because people in these countries have more wealth and resources at their disposal, and seemingly could have and raise children much more easily.

Part of the reason behind this demographic trend is an economic one. While people in wealthier countries make more money annually, their resource consumption and cost of living are much greater. A typical family in a developing country, for example, may be greatly concerned about having enough food to feed everyone adequately while maintaining shelter from the elements. In a developed country, however, parents are expected to provide transportation, entertainment, and post-secondary education opportunities. The cost of raising a child is much greater in these countries.

As with developing countries, there are also some sociological factors at play. Developed countries have a much greater rate of antinatalist forces—variables that decrease one's likelihood of having children.

Antinatalist forces all revolve around women. Specifically, it has been observed that women who have more opportunities, more equal say in family decisions, and a higher level of education are less likely to have large families.

# The Demographic Transition Model

We've defined two ends of the demographic spectrum: the developed and developing worlds. However, the picture is not quite this simple. Countries are not necessarily defined as being in one camp or the other; there is a lot of transition area in between.

The economic and demographic fortunes of a country are not written in stone. All countries have been observed to pass through a series of four stages, each marked by major differences in economic growth, and each with a different population picture.

This gradual evolution of a country is described as the demographic transition model. Beginning with conditions consistent with a developing country, living conditions, culture, values, and traditions that relate to population growth all undergo an evolution. The end stage more closely resembles a wealthy, educated, developed country.

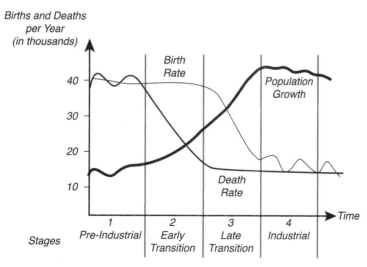

*The demographic transition model, a pattern of population growth and changes in birth and death rates seen as countries industrialize.*

## Stage One: Pre-Industrial

With each stage in the demographic transition, we will track three demographic variables: birth rate, death rate, and overall population growth. A country in the pre-industrial stage is experiencing a high birth rate and a high death rate. The overall population growth rate of such a country is stable. The birth rate and death rate essentially cancel each other out.

Why is the death rate so high? Countries in this stage don't have access to many of the medical and technological advances that have lengthened human life expectancy. Water treatment, antibiotics, and vaccinations may simply not be widely available. There may be other health issues as well, such as famine or malnutrition. In a country like this, high birth rates are a necessity simply to maintain the population.

Married couples in pre-industrial countries probably aren't having morbid conversations about how many children they need to have to ensure that, given the infant mortality rate, two or three survive to adulthood. However, there may be many cultural pressures to have lots of children. The dominant religion may actually forbid any form of birth control, assuming it is even available.

The population pyramid for Ethiopia we looked at earlier is actually reflective of a pre-industrial country. The base is very wide, indicating high numbers of children born every year. The top of the pyramid is very narrow, indicating few of those children are surviving to advanced age. No countries today really fit the exact definition of pre-industrial, as most have at least some access to modern health care. Even Ethiopia is experiencing a net population growth.

## Stage Two: Transition

In the transitional stage, one big demographic change occurs: the death rate drops. The introduction of even modest improvements in water sanitation or medical care in a country will have huge effects on the death rate.

While it is relatively easy to increase life span and decrease the death rate in a country through foreign aid or economic growth, changing the birth rate is a much different challenge. After all, what is the basis of most of the pronatalist forces that encourage birth rate? Tradition and culture, two societal influences that change very slowly.

As a result, we have a now-lowered death rate, but a birth rate that stubbornly remains high. This differential causes the population growth rate to accelerate very quickly. Ethiopia, our example from before, is more accurately placed in the early part of the transition stage, because they are experiencing an annual growth rate of 2.1 percent (doubling time = 33 years).

The transition stage is named so because the country is undergoing change. Death rate declines first. As the country progresses through the later part of this stage, birth rates will start to drop as well. What is the impetus for this change? This varies from country to country, but usually it involves an acceptance of birth control, increased rights and opportunities for women, and an overall more highly educated populace.

What would a late transition country's population pyramid look like? The proportion of the youngest age groups starts to shrink, reflecting a conscious decision on the part of the adults to have smaller families.

India has been making efforts for decades to provide birth control education and promote equal educational opportunities for women. Consequently, India's annual growth rate is currently 1.4 percent (doubling time = 50 years). While this is good news, the growth rate is dropping very slowly. By the time India's population comes into equilibrium, its may surpass even China's.

## Stage 3: Industrialized

The industrial stage is marked by a return to a balance between birth rates and death rates. Once a country has reached this stage, the population growth rate is close to zero.

At this point in the demographic transition, the population explosion of the transitional stage is over. However, by the time this has been achieved, the population may have doubled, tripled, or more.

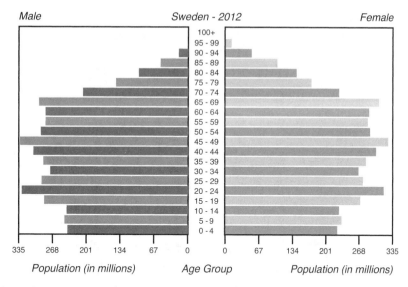

*The population pyramid for Sweden as of 2012, according to the U.S. Census international database.*

Industrial stage countries have a population pyramid that actually more resembles a rectangle. The different age groups are pretty comparable in size. This reflects a replacement-level family size (2.1 kids per couple on average) in the adult population, and adequate health care and sanitation to allow most adults to reach their life expectancy limits.

 **A LOOK BACK**

Following the end of World War II, the United States experienced a baby boom. The total fertility rate peaked at over 3.5 children per woman. Through the 1960s and 1970s, this trend reversed, resulting in a baby bust and a total fertility rate below replacement level! These two decades saw a huge societal change in the United States. Women began pursuing advanced degrees and entered the workplace in greater numbers, and birth control became more accepted. The 1970s also saw the *Roe* v. *Wade* Supreme Court decision legalizing abortion. While the birth rate in the United States has since returned to replacement level, the rate of transition to the industrial stage was an extraordinarily rapid one.

# Stage Four: Post-Industrial

In theory, the industrial stage should be the end of the demographic transition. Birth rates and death rates are at equilibrium, and the population is stable and maintained.

However, in some developed countries, the trend of declining birth rates continues well below the replacement level birth rate. Facing a competitive work environment, women may put off having children until later ages or choose to not have any at all. An elevated cost of living may put pressure on the members of the household to have fewer children.

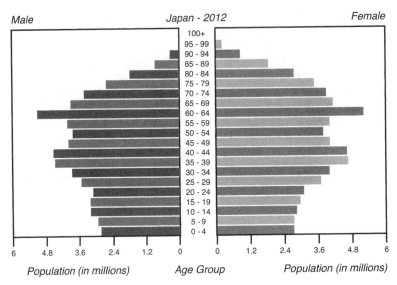

*The population pyramid for Japan as of 2012, according to the U.S. Census international database.*

An illustrative example of this phenomenon is Japan. Japan is experiencing a shrinking population, as their fertility rate has dropped below replacement level and immigration is discouraged. The population pyramid for Japan actually looks inverted, as the older age groups outnumber the young.

The human population is clearly growing, but this isn't a hopeless situation. Birth rates in many developed countries are declining, a result of greater lifestyle and educational choices among women. As these cultural changes work their way into the developing world, the human population will eventually stabilize.

## The Least You Need to Know

- The human population grew roughly at a linear rate up until the Industrial Revolution, when exponential growth began.
- Developing countries are the poorest with the most difficult living conditions, yet have much higher birth rates than developed countries.

- Factors that decrease one's likelihood of having children primarily revolve around educational and professional opportunities for women.

- The demographic transition is a series of stages that countries pass through as their death rates fall due to improved living conditions, and birth rates eventually follow.

# CHAPTER 10

# Environmental Economics

Economics might seem like a completely foreign field of study for an environmental scientist. However, remember that environmental science is interdisciplinary. Many of the decisions regarding how we use resources or handle problems of disease and pollution are rooted in economics.

Why are so many consumer products available in developed countries made in the developing world? Why is coal relied upon so heavily as a source of electricity when its polluting effects are well documented? Why are the benthic zones of many of the world's lakes and seas coated with nondegradable pollutants from decades ago?

The answer to each of these questions involves economic costs. Money is a huge driving factor in our short-term and long-term decision making as a society. Many of these decisions have environmental consequences, both seen and unseen.

## In This Chapter

- The relationship between economics and societal decision making
- The hidden costs of producing goods
- The ecological impact of a society
- Economic laws and regulations that promote sustainability

# Money and Resources

Before we can closely examine the relationship between the environment and economics, there are some basic rules and relationships you must understand.

While each of these ideas is rooted in the study of economics, many of them have parallels in environmental science. These basic principles govern the economic growth of our society, assign value to specific goods and services, and help us prioritize different aspects of a problem when making a decision.

## The Earth's Investments

In terms of wealth, capital is considered the amount of money or goods that can be used to produce more wealth. For example, a bank may have millions of dollars of assets that it holds from the collective of all its customers.

This money can be loaned out to other individuals so they can buy additional goods and services or start their own business. The bank will charge a set amount of interest on this loan, money that the borrower will pay back in addition to the actual amount of the loan. Assuming all goes well, the bank should profit from this arrangement.

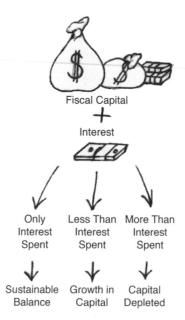

*A financial institution that spends only the interest it receives or less will be sustainable. There will be no net loss in capital. Spending more will gradually deplete this capital.*

In the environment, *natural capital* is the sum of all the resources available for use by our society. This could include anything from clean water to timber, from coal to fertile soil. In ecology, there is also *natural interest*. For example, let's consider the entire stock of tuna in the ocean. This would be our natural capital. Each year, a certain number of new tuna will be born into this population. This would represent the natural interest.

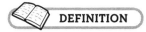 **DEFINITION**

> **Natural capital** refers to all the resources available from the Earth, everything from fish to timber to clean water. Over time, a given part of the Earth will grow some of these resources, such as when a fish population increases. This is **natural interest.**

Each of these systems works as long as a certain amount of capital is always left behind for future use. If the bank were to loan out or spend all its money, it wouldn't have enough to run its daily operations or provide cash to its customers. Similarly, if the tuna populations of the oceans are continually overharvested, the incoming interest will not be enough to offset the loss. Either way, we eventually exhaust our capital and are left with nothing.

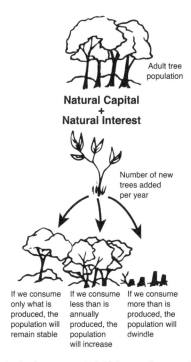

*Considering the adult trees in the forest as capital, if they are harvested at a greater rate than their saplings can regrow, then the forest itself will be depleted.*

## Supply and Demand

Natural resources are not all consumed at the same rate. Some examples of natural capital are relatively inexpensive. For example, in the United States clean water costs the average household little. Other resources, such as the crude oil that is eventually refined into gasoline, have become increasingly expensive over the last several years.

The relative cost of each of these resources is primarily the result of supply and demand. The price of a resource will increase if its supply is low or its demand is high. Sticking with our original example, demand for both oil and water is high. Most people in developed countries are dependent on vehicles for transportation, and everyone needs clean water for drinking, cooking, and bathing.

The difference between the two resources is in their supply. Most areas of the United States are pretty plentiful in terms of freshwater availability. Some states are adjacent to the Great Lakes, others have large rivers passing through, and still others are able to pump their water from underground.

 **A LOOK BACK**

Peak oil is a term that describes the point at which a region has reached its maximum potential for drilling and extracting crude oil. In the case of the United States, this occurred in the 1970s. In 1973, to pressure Western countries to cease their support of Israel in a regional conflict, the oil-producing countries of the Middle East instituted an embargo. The market price quadrupled, and the United States was forced to try to find ways to drastically reduce oil consumption and research renewable energy such as solar and wind power. When the prices eventually eased, many of these conservation efforts went by the wayside.

Some resources are relatively easy to extract; others are much more difficult. Gold, for example, is almost never mined in the old-fashioned way with pickaxes and sifting pans anymore. Rather, huge dump trucks deliver dirt and ore to a facility where liquid cyanide is used to extract the gold. This is an extremely difficult, expensive, and dangerous process, but economically worthwhile. The price of gold is currently over $1,500 an ounce! The demand for it is very high, and the overall supply is low, especially as the largest and most accessible deposits have already been mined.

## A Society's Footprint

The ecological footprint of a society includes all the land needed to provide its resources and store its waste. As an example, the United States has an ecological footprint of over 20 acres per person. Multiplied across the entire population of about 300 million people, this works out to about 6 billion acres of land needed to support the U.S. population. Compare this to the actual land space of the United States, which is about 2 billion acres. As a comparison, China, with a much greater population (1.3 billion), uses only 2 acres per person.

Why does the United States have an overall larger footprint? One model that demonstrates the relevant variables is called IPAT. The model works under the premise of a simple equation:

$$\text{I (Impact)} = \text{P (Population)} \times \text{A (Affluence)} \times \text{T (Technology)}$$

Simply put, this model suggests that the ecological impact of a society is directly proportionate to its population size, wealth, and level of technology dependence.

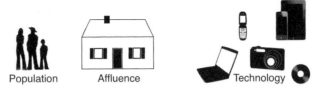

*The three main variables that influence the ecological impact (I) of a society are the population (P) size, level of affluence (A), and available technology (T).*

An increased population directly affects the ecological impact of a society, as more land and resources will be needed, and a greater amount of pollution will be produced. An affluent society will likely produce more goods to meet the demands of its people, which in turn will increase the ecological impact.

For example, an article by Kouikoglou and Phillis in *Ecological Economics* estimated that the construction of one car had the following impacts, among others, on the environment:

- Over 600,000 gallons of water to manufacture parts and tires.

- Nearly 700 pounds of mine waste from extracting the lead needed for the battery.

How does technology factor into this? Consider the forms of resource extraction that are available now that weren't just a few decades ago. Oil can be drilled from the bottom of the ocean, extracted from the tar sands of Canada, and obtained from many other places that were formerly unreachable. Increased mining, refining, and the eventual use of the oil all increase the environmental impacts of the society.

**CASE STUDY**

Rare earth metals are a collection of elements that are found naturally in the Earth's crust. These are very important minerals in the manufacture of many consumer electronics. These elements aren't actually all that rare, but they exist in very small quantities throughout the Earth's crust. Enormous amounts of dirt and rock must be mined, transported, processed, and separated just to get a small amount of the metals. This is an incredibly complex and expensive process, but the demand for rare earth metals makes it a lucrative business.

# Money and Politics

One might conclude that if the principles of economics were the only ones to govern environmental decision making, short-term benefits would be the biggest priority. Historically, this has often been the case.

In 1969, a pesticide factory was built in the city of Bhopal, India. The factory produced Sevin, a brand-name insect poison that kills a wide range of pests, including mosquitoes. As the plant operated into the 1970s and early 1980s, small leaks of various chemicals were commonly reported. The local authorities warned Union Carbide, who owned the plant, about these issues, but no action was taken. In 1984, an explosion at the plant leaked a massive plume of these chemicals into the air, which drifted over the nearby city, causing thousands of deaths and hundreds of thousands of injuries and illnesses.

The explosion occurred as a result of the plant not being properly maintained. Several safety systems had failed, and others were in disrepair. Emergency action plans were inadequate, and the workers were not properly trained. Did this disaster have to happen?

## Does the Cost Outweigh the Benefit?

In hindsight, this tragedy was largely preventable if the company had spent enough money to maintain the plant, train the workers, and develop emergency containment procedures. And yet they failed to take these precautionary measures. Why?

At some point, the company determined that the cost of repairs, maintenance, and training outweighed their perceived benefits. This decision-making process is called a cost-benefit analysis. The costs were short-term investments in the plant. The benefit was a long-term reduction in risk of an accident. The company clearly felt either that an accident was very unlikely or that if an accident were to occur, it would be manageable.

Decisions are often made using this model. It can be effective, assuming that the short-term and long-term costs and risks are estimated and calculated accurately. However, the costs and risks of an action can be difficult to accurately predict. In the case of the Bhopal plant explosion, the short-term investment cost in the plant was clear—a set dollar amount was needed. The long-term risks were not clear. There had been leaks at the plant before, but they had always been very small, affecting only a few workers at a time.

## Visible and Hidden Costs

With the pesticide plant in Bhopal, there were two levels of costs involved in the production of the chemicals. The internal costs reflected the money needed for the actual day-to-day operations of the plant. Materials, salaries for workers, and energy for heat and electricity are all internal costs. These costs are reflected in the price of the final product. If the cost of materials goes up, the final price paid by the consumer does as well.

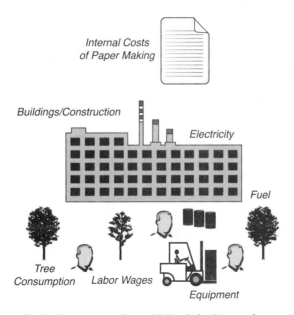

*The internal costs of producing paper are those paid directly by the manufacturer. The price of the paper reflects these costs.*

There are other costs as well, ones that are not directly part of the price equation. Environmental damage caused by the leaking chemicals, as well as health effects in Bhopal as a result of these chemicals, are external costs. These are not typically reflected in the price of the final goods, because they are often not directly paid by the company.

Imagine that these costs were actually reflected in the final price tag of the product. This would be the true cost of the pesticide. Unfortunately, the true cost is rarely used in this way. As a result, the public purchases products at a certain level of demand based on the lower price tag.

The demand, now elevated, results in the production of even more of the product, and the hidden external costs continue to mount.

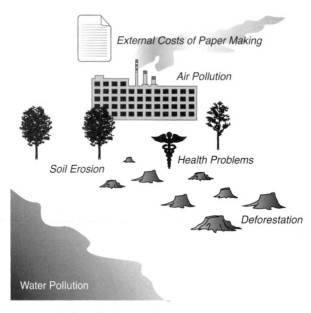

*External costs are not directly borne by the manufacturer; rather, they affect the environment and the society as a whole. The price of the product does not reflect these costs.*

A striking example of these costs can be seen in a simple consumer item: a can of tuna. A standard brand-name can of tuna at the grocery store is priced between $1.00 and $2.00, depending on whether there is a sale. Higher-end grocery stores may also carry another brand of tuna that is priced closer to $4.00 to $5.00 per can. The difference between these two is that the latter factors into hidden external costs.

## External Cost Comparison in Two Brands of Tuna

|  | Major Brand | Local Brand |
|---|---|---|
| Consumer price | $2.00 | $5.00 |
| Source | Indonesia | Western United States |
| Fair wage for workers | No | Yes |
| Working conditions | Mostly unregulated | Regulated |
| Fished sustainably | No | Yes |

The cost of this simple food staple grows significantly when the factory workers are given a fair, livable wage, when working conditions are regulated and monitored, and when sustainable fishing practices are followed. The cheaper tuna has a lower price not because the costs are less, but because fewer of those costs are internal; they are borne instead by the workers, the populations of tuna, and the ocean ecosystem itself. An obvious question remains, then: why are the minimum wage and working condition laws and regulations not present in other countries like Indonesia?

## The Price of Regulation

These regulations are not without cost. The ability of a company to grow, expand, produce, hire, and ultimately maximize profits is hindered in some way by regulation. A regulated coal mine forced to equip all their workers with respirators and poisonous gas detectors will have a reduction in profits compared to one that is not regulated. With increased profits, the second coal mine may expand at a greater rate and hire more workers. This will reduce the unemployment rate of the surrounding area and result in an increased level of economic prosperity.

 **A LOOK BACK**

At the time it was constructed, the Hoover Dam was the largest public works project ever attempted in the United States. However, if the dam were proposed today, it likely would not be constructed.

While the dam provided many benefits, such as a stable reservoir of water, flood control, and hydroelectricity, it also had a huge ecological impact on the Colorado River. Over 100 workers died during its construction. Modern worker safety requirements and environmental impact regulations would have elevated the price tag far above the $49 million final cost ($750 million in today's dollars).

Does this mean that regulation is economically a bad thing? That's a difficult question to answer, but in the short term, it often does have the effect of slowing down economic growth. Remember, though, we aren't just considering short-term benefit. We have to look at long-term consequences and hidden costs, too.

## The Benefits of Regulation

Equipping coal workers with respirators, installing poisonous gas detectors, and employing other safety measures would certainly result in greater overall profits for the company. To take it a step further, what if the workers are not given any health benefits? What if their wages are kept so low that they all have to live in the immediate area around the mine? What if the mine is allowed to dump its waste into a nearby riverbed?

At some point, long-term costs will begin mounting for this operation. The workers eventually will suffer from health ailments, such as black lung disease. Without health insurance, their care will be completely the responsibility of their families. Family members, now forced to care for these physically disabled workers, may not be able to get a job themselves. Meanwhile, the primary water source for the community has become increasingly toxic as pollutants in the ore from the mine have leached into the water.

The initial economic benefits of the mine have been outstripped by the long-term costs of not having any regulations. The problems described above will likely arise with any coal-mining operation. However, with regulations in place governing worker safety, health care, and water quality, the magnitude of these problems will be much more manageable. After all, in hindsight, don't you think the people of Bhopal and the owners of the Union Carbide pesticide factory would have gladly supported better maintenance of the facility if it could have averted that tragedy?

# Money and Environmental Ethics

In a sense, the consumer may seem directly responsible for many of the environmental problems that have resulted from these economically centered decisions. After all, a tuna company couldn't operate unsustainably and make unfair demands of its workers if it didn't have any customers!

To an extent, it's fair to bestow a certain amount of responsibility on the consumer. However, in many cases the ordinary consumer may be completely unaware of the hidden costs behind the products they're purchasing. When you go to the grocery store, do you know exactly how each of the products was made, where it was made, and the environmental effects of its production? This is an unfair expectation, but there are actions that can be taken to strike a balance and create a more environmentally sustainable economy.

## Eco-Labeling

Over the last several years, a movement has emerged to begin labeling certain products based on the concern behind the hidden costs of making it. As a result, a handful of certifications have emerged to address these concerns. A consumer familiar with these different certifications can make purchasing decisions to support firms that operate in a more responsible and sustainable manner.

Fair trade certification is a system that audits producers of imported goods to see if they're meeting minimum standards of worker compensation and treatment. Examples include setting minimum livable wages and allowing farmers to collectively bargain contracts. This program is not free from criticism, however, as the amount that actually reaches the farmers is not consistent.

Factories themselves can also be certified. *LEED* is an internationally recognized green building certification system. Practices such as reducing industrial waste, recycling materials, and minimizing the use of toxic materials are all part of the certification process.

The *Marine Stewardship Council* (*MSC*) has a certification for fisheries that use sustainable harvesting practices. Requirements for this certification include limiting annual catches so the fish populations can replenish themselves. MSC-certified seafoods are typically more costly to the consumer, but are much healthier to the greater ocean ecosystems in the long term. There is also a *Forest Stewardship Council* (*FSC*) certification that recognizes sustainable practices in the manufacture of wood products.

 **DEFINITION**

> **Fair trade** is a label given to brands that meet standards promoting sustainable practices and providing fair compensation. **MSC certification** recognizes fisheries that agree to limit their annual harvests of fish to more sustainable levels. **LEED certification** carries a set of standards that new buildings can meet involving energy efficiency and use of green materials. **Forest Stewardship Council** (**FSC**) certification is given to manufacturers of paper and wood products that meet standards of sustainability and fair trade.

Money is a huge driving factor in the way our society operates. Due to this influence, many decisions are made to maximize short-term prosperity at the expense of long-term sustainability. While some economic regulations help to ease this strain, many goods we buy still carry a hefty hidden cost to the environment.

## The Least You Need to Know

- Goods and services that are either high in demand or low in supply will have the greatest price.
- Each product has a set of costs, internal and external, that are paid either directly by the consumer or by the society and environment.
- Societies with larger populations, greater affluence, and technological advancements have the greatest impact on the environment.
- Eco-labels such as fair trade are certifications that companies can achieve to advertise themselves as more environmentally conscious.

# CHAPTER 11

# Plant Agriculture

Humans are omnivores—we eat both animal and plant matter. We're also tertiary consumers, meaning we're at the top of most food chains. Normally, in any given ecosystem, the number of secondary and tertiary consumers would be pretty low. There simply isn't enough energy in the ecosystem to support more. Yet the human population is more than 7 billion and growing.

Our burgeoning population and dietary requirements present a huge challenge. How do we continue to provide adequate nutrition to the entire human population?

Toward the end of the last glacial period, the human race transitioned from a hunter-gatherer society to an agricultural one. We began producing our own food, including breeding species of animals and plants specifically for our needs. This ability to cultivate other species for our own use has enabled our population to grow into the billions.

However, farming now is much different than it was in the early days of civilization. Many technological advances in plant and animal breeding, genetic manipulation, and soil cultivation have been essential to keep up with our population growth. Many of these advances have brought unforeseen consequences.

## In This Chapter

* The basic human nutritional needs
* How the Green Revolution changed farming
* The composition of soil
* How soil is degraded by human activities
* Organic versus genetically modified crops

# Our Nutritional Needs

Before we can examine the inner workings of farming and agriculture, we need to understand exactly what nutrients humans need to survive.

In general, people need about a thousand calories a day simply to survive. This is not taking into account whether the person has an especially active lifestyle, nor does it consider any micronutrients like vitamins and minerals. These calories and nutrients can come from a wide variety of sources, and the diets people subsist on vary widely from region to region.

## The Big Nutrients

As we covered before, there are three main sources of energy in living organisms: carbohydrates, lipids, and proteins. Each nutrient has a different set of sources it can be found in, and each is essential to our bodies in a different way.

### Macronutrients

| Energy | Yield | Function |
|---|---|---|
| Carbohydrates | 4 kcal/g | Primary energy source |
| Lipids | 9 kcal/g | Forms cell membranes |
| Proteins | 4 kcal/g | Enzymes and body structures |

While each of these sources does provide energy, a healthy diet consists of a mixture of all three. There are some molecules the human body is not able to produce on its own, and these must be included in the diet. Any diet that's lacking in one or more specific nutrients can result in serious health effects. This condition is called malnutrition, and there are many different types prevalent throughout the world.

## Food Staples

Wheat, corn, and rice make up the vast majority of food produced in the world. In fact, most of the food staples the world population depends on are plant based. You have to move all the way down to the sixth ranking to see an animal product, fish.

Why are plants relied upon so heavily as a source of energy compared to animal sources? The answer actually lies within two ecological concepts: the energy pyramid and the 10 percent rule. Remember, producers make up the base of every food web, as they directly absorb energy from the sun. As you move up a food chain, the available energy decreases. Only about 10 percent of the available energy makes it up to the next level; the rest is lost as heat. How does this relate to human food production? The answer is simple: growing plants and directly consuming their energy is more efficient!

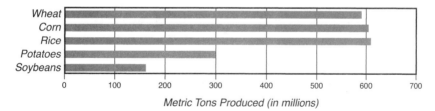

Metric Tons Produced (in millions)

*The top five crops produced around the world, according to the United Nations Food and Agriculture Organization.*

## Missing Nutrients

Unfortunately, meeting human nutritional needs isn't as simple as providing a certain number of calories per day. If this were the case, entire populations could sustain themselves on nothing but rice.

Human dietary needs are complex when it comes to the micronutrients—vitamins and minerals. When any of these nutrients are absent from the diet, malnutrition occurs. The exact symptoms and signs vary, depending on what nutrient is deficient.

One of the most common deficiencies in the world is iron. Iron is an element found in meat and leafy green vegetables such as spinach. A single atom of iron is present in the hemoglobin (oxygen transporter) molecule in each red blood cell. If iron is not available, the hemoglobin, and thus the red blood cell, cannot be made. An insufficient number of red blood cells results in decreased oxygen flow to numerous areas of the body, most importantly the brain. This is called anemia. This condition could be prevented entirely if everyone on Earth had an adequate supply of iron. However, many populations are restricted to just a few plant-based food staples that are low in iron.

**CASE STUDY**

Iodine is an element that occurs naturally in some soils of the world. It's incorporated into plants and used by the human body to synthesize hormones produced by the thyroid gland that control body metabolism. People in regions with iodine-poor soils and no access to alternative sources like seafood are prone to iodine deficiency, which causes an enlargement of the thyroid gland, called a goiter. This disorder is almost unheard of in the developed world, because iodine is added to table salt. Iodine deficiency is considered the easiest and cheapest of all disorders to prevent, and efforts are underway to provide supplements to regions that need it.

## Missing Calories

The most fundamental form of malnutrition is an absence of sufficient calories. This condition is actually called undernutrition. Remember, the minimum amount for survival is about 1,000 calories per day, and in many regions of the world, this amount may not be available to all.

In theory, there should be enough calories in the food produced on Earth to provide more than enough for everyone. However, this is assuming that no food is wasted and it's fairly evenly distributed. Neither of these is the case. Chronic hunger is a problem especially prevalent in developing countries.

What causes famines, or large-scale undernourishment of entire populations? The single biggest cause is actually drought. Severe droughts have occurred throughout history, such as the ones that contributed to the collapse of the Mayan civilization during the ninth and tenth centuries. Without a sufficient supply of water, no crop will grow, and if food cannot be imported from elsewhere, famine is inevitable.

One of the greatest famines in recent history occurred in North Korea during the 1990s. The famine peaked due to severe flooding that destroyed farmland and harvests throughout the country, especially in 1995. The country didn't have the infrastructure or governance to handle such a crisis, and was even reluctant to accept foreign aid until the floods of 1995.

# The Green Revolution

A potential answer to famines in the developing world emerged in 1961, as India was facing a massive famine. A scientist named Norman Borlaug, well known for his work in modernizing agriculture in Mexico, was asked to introduce high-yielding strains of wheat he had cultivated.

After a few trial runs, full planting began in the food-producing regions of India. Special attention was given to the Punjab region, India's breadbasket and one of the most fertile places on Earth.

The yields of the new wheat strains were the highest of any crop planted in Southeast Asia. Huge amounts of the seed were imported during the following years, and wheat yields in both India and Pakistan doubled. Borlaug is credited with saving the lives of millions of people worldwide by increasing food supply in countries like these. He won the Nobel Peace Prize in 1970 for these efforts, labeled the "Green Revolution."

Cultivated Wheat

Wild Wheat

*Modern cultivated wheat plants look shorter, but have much larger grains with a greater amount of starch.*

## The Recipe of Farming

Changing the strains of wheat and other crops planted in the region was actually only one part of the Green Revolution. Equally important was increasing the agricultural inputs, added measures given to the soil to increase the growth of crops.

The first input is water. Prior to the Green Revolution, most food-producing areas depended strictly on rainwater. This leaves the farmer vulnerable to seasonal changes (many regions have wet and dry seasons) as well as the occasional drought. Irrigation, adding water from external sources such as wells, rivers, or lakes, helps to alleviate this uncertainty. During this revolution in agriculture, canals were dug, dams were constructed, and reservoirs were filled to improve irrigation.

Plant growth is also limited by the availability of certain nutrients in the soil. The most important plant nutrient is nitrogen, followed by phosphorus and potassium. The Green Revolution introduced the use of synthetic fertilizers, much like the kinds you can get from the hardware store today.

The use of machines can also greatly improve efficiency and yield. Plowing, harvesting, and many other farm labors that had previously been the domain of large animals or humans now could be completed more quickly by tractors and combines.

## The War of the Pests

The last great limiting factor in maximizing crop yields was pests. Insects, rodents, and birds would eat the seeds, the leaves, or the crops themselves. Weeds would grow among the cultivated plants, outcompeting them for precious water and soil nutrients.

The Green Revolution also ushered in a new era of synthetic pesticide and herbicide use. These chemicals could be sprayed in large doses over entire fields, selectively killing the pests and weeds while leaving the plants alone.

These chemicals were not without their side effects. While they were very effective at eliminating pest animals and plants, they often had the side effect of killing beneficial organisms, such as earthworms. There was also concern about the residues from these pesticides remaining behind in the crops after they were harvested. However, these concerns were largely put aside for the time being, as the increase in crop yield was simply too great and the need for higher food production too important.

## Genetically Modified Organisms

The Green Revolution seemed to have accomplished something monumental—the carrying capacity of the Earth increased. Scientists and demographers had been increasingly alarmed at the population growth of the twentieth century and foresaw the deaths of billions from famine. This disaster had largely been averted, yet the human population continued to grow exponentially, pushing the limits of crop production even further.

Human innovation continued with the introduction of specific genes from unrelated species into food crops. This technology meant that a specific desirable trait from one organism could be transplanted into another. While humans have been hybridizing and breeding domestic plants for millennia, this technology meant the introduction of new breeds and strains at a rate never before seen.

 **CASE STUDY**

Golden Rice is a genetically modified strain of rice. Plants with the inserted genes have the ability to produce significant amounts of vitamin A, which is normally completely absent from the grain. Populations that are dependent on the crop are very susceptible to vitamin A deficiency, a leading cause of blindness in these societies. Golden Rice could, in theory, alleviate this form of malnutrition. However, the grain is not without problems. Initial strains didn't produce sufficient vitamin A to be a supplement. Furthermore, this vitamin requires some fat in the diet to be absorbed properly. Finally, opponents argue that Golden Rice treats one of the symptoms of poverty, but not the cause.

There are dozens of genetically modified organisms. We'll focus on three widely used examples, all found in corn. The first is called Roundup Ready corn. This is a strain that has the ability to resist an herbicide that effectively kills all other plants. The use of this plant means that an entire field can be sprayed with the herbicide, killing every plant except for the desired crop.

Bt corn has a gene from a species of bacterium, *Bacillus thuringiensis* (hence the Bt moniker). These bacteria produce a toxin that's poisonous to specific insect pests, primarily the European corn borer. Bt corn containing this gene can produce the toxin independently, removing the need to spray the plant at all!

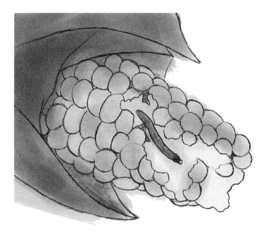

*The corn borer is a very destructive pest that chews tunnels through the stalks and ears of corn plants, causing them to fall over.*

Genetically modified organisms are not without criticism. Like the use of pesticides and herbicides, there's potential for accidentally affecting other species. Concern has also been raised that these crops could cause an increase in food allergies due to the proteins produced by the plant (like the Bt toxin) that normally wouldn't be present. Finally, the notion that a single corporation, such as Monsanto, has the ability to patent and control the distribution of seed is a completely new idea that many find discomforting. Farmers who use Roundup Ready corn are not allowed to re-use seed from their last harvest but must purchase a new batch every season.

# Soil, the Foundation of Food

The increased inputs of the agricultural revolution had the effect of raising crop yields. Part of this is due to the plant actually growing faster and larger, and producing more fruit or seed. Farmers also are able to grow plants more intensively—they can plant greater amounts of the same crop on a given parcel of land.

While the improved crop yields are unquestionably good for reducing famine, a tremendous strain is placed on the soil. Soil is a complex mixture of organic matter and minerals, and no plant can grow without it.

# A Complex Mixture

Soil is a complex mixture of matter that has been broken down over time. The parent materials could include rock or smaller sediments brought by wind, water, or glacier movement. The forces of nature gradually wear down these larger materials into smaller and smaller fragments.

Eventually, these fragments are reclassified as one of three different types, based on size. Sand particles are the largest, followed by silt, and then clay. Consider the following example. An amount of sand is dropped into a container of water. The sand, composed of larger, heavier particles, sinks to the bottom. A separate container of water is mixed with clay. The clay, composed of lighter, finer particles, mixes and becomes suspended in the water. This creates the cloudy appearance people typically associate with dirty or muddy water!

Most soils are composed of different combinations of sand, silt, and clay. These mixtures are called loam. Scientists can perform a simple test to determine soil composition by mixing a soil sample with water and watching how long it takes the different materials to settle. Remember, each has a different size and mass.

## Soil Texture and Plant Agriculture

| Soil Texture | Nutrient Retention | Moisture Retention | Ease of Use |
|---|---|---|---|
| Sand | Poor | Poor | Easy |
| Silt | Medium | Medium | Medium |
| Clay | Good | Good | Difficult |
| Loam | Medium | Medium | Medium |

As you can see from the table, each type of soil texture has upsides and downsides for farmers. Sand is very easy to work with, but it doesn't hold water and nutrients well. Clay is much better at retaining water and nutrients, but it is extremely difficult to work with. Most farmers prefer loam.

# Layers of Soil

When you think of soil, images of loose black or dark brown dirt intermingled with earthworms and other insects probably come to mind. This is actually only one layer—the topsoil. Other layers are found underneath, each with its own unique characteristics. Understanding these layers of soil is important, because it sheds light on why certain areas are so fertile while others support only limited farming.

Technically, topsoil isn't the layer closest to the surface. The first layer is actually called the O Horizon. This name comes from its being all organic matter: leaf litter, plant roots, insects, droppings from birds and animals, and so on. The topsoil is actually called the A Horizon, and has the highest percentage of organic matter of any layer. Topsoil receives all the organic matter from the O Horizon once it has been fully decomposed. This is what produces the dark brown or black color.

Beneath the organic litter and topsoil is the subsoil, also called the B Horizon. Subsoil looks and feels much different from topsoil. It's paler in color and may carry a tint of the minerals it contains, such as red from iron. Subsoil has very little organic matter, instead being primarily composed of minerals that leached downward (from absorbed rainfall) out of the A Horizon.

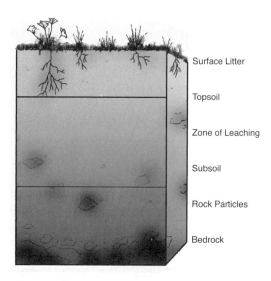

Surface Litter

Topsoil

Zone of Leaching

Subsoil

Rock Particles

Bedrock

*Plant roots and living organisms are typically found only through the topsoil layer. The subsoil and parent rock beneath are made of inorganic minerals.*

The layers below the topsoil can hardly be considered soil, as they're primarily rock. The C Horizon is referred to as "parent rock" because it's mostly large, unweathered rocks. Farther below is the R Horizon, which is pure, unbroken stone, also called bedrock.

In terms of plant agriculture, ecosystems with deeper layers of organic litter and topsoil are considered the most fertile. Topsoil contains a good mixture of minerals, is loose enough for plants to grow roots into easily, and holds moisture from rainfall well. Subsoil doesn't support as much growth, as it lacks organic matter.

## How Plants Respond to Soil

The primary goal of all farmers is to have healthy plants that lead to the biggest crop yield possible. Plants all need the same basic things to grow: ample sunlight, appropriate temperature levels, air, nutrients, and water. The soil has a huge influence over the last three variables on that list. What are the characteristics of fertile soil?

Plants need mineral nutrients to grow, including phosphorus, potassium, and calcium. These minerals are added to the soil to some extent by the weathering of rocks, but this process takes much too long. The primary source of these nutrients is actually the decomposition of dead plant and animal matter in the O Horizon. Farmers supplement this with fertilizer. However, no amount of fertilizer will help much if the soil is unable to retain these minerals until the roots take them up. This is called the nutrient-holding capacity.

Water is also extremely important to plant growth. Water is constantly being absorbed into the roots, transported through the stems, and eventually transpired (evaporated) through the leaves as the plant breathes. Plants that don't receive sufficient water will wilt, shutting down photosynthesis. Ideally, the soil will allow water to easily infiltrate, or soak in. Once the water is absorbed, it must be retained by the soil long enough for the roots to use it. This is called the water-holding capacity.

# The Degradation of Soil

In a natural ecosystem, the nutrients and minerals in a soil are continually replenished through the biogeochemical cycles. Animals and plants die, decomposers break down the remains, and the compounds within are released into the air or soil. When humans enter the picture, many of these processes are disrupted, leading to soil degradation.

One of the most illustrative examples of how devastating soil degradation can be is found in Haiti. This country occupies the western side of the island of Hispaniola, so named by Christopher Columbus because its lush green forests reminded him of Spain. When the island was occupied by Spanish and later French colonists, much of the native population was wiped out. African slaves were imported to carry out the labor of farming and mining.

A slave revolt in the late eighteenth century ended the rule of the French. The land was divided up among the now-freed slaves. As the Haitian population grew, pressure on the agricultural system mounted. Farmers began overplanting, which quickly stripped the soil of nutrients. Forests were leveled for timber and charcoal, and to open up more farmland. The soil, no longer protected by the roots of the native plants, became highly susceptible to mudslides during heavy tropical rains. Today, less than three percent of the original forest cover remains.

## Nutrient Depletion

Nutrient depletion is simply when the minerals needed for plant growth are removed from the soil faster than they are replaced. This has the potential to happen whenever intensive farming occurs. One of the reasons the Green Revolution was so incredibly successful in raising crop yields is that it replaced these nutrients with synthetically produced ones.

Nutrient depletion is actually one of the key reasons why rainforest soils are such poor choices for farming. This seems counterintuitive, as they support a vast amount of plant life. However, the rainforest plants grow so fast that the decomposers on the forest floor (which also grow at an extraordinary rate) are barely able to keep up. When a farmer takes over a patch of forest and burns or clear-cuts it, the nutrients in the soil are typically used up very quickly.

With the introduction of synthetic fertilizers, this problem should largely be solved—right? Unfortunately, this may not be the case. A USDA study in 1999 showed that the levels of certain mineral nutrients in crops, like calcium, were significantly lower than measurements taken during the 1960s. Calcium levels in broccoli, for example, were found to be over 50 percent lower. This may be an unintended consequence of the Green Revolution.

## Soil Loss Through Erosion

Soil as we know it wouldn't exist without *erosion*. The action of wind and water breaks down large rock into smaller fragments, eventually forming soil. Erosion is a natural process that has been occurring throughout the Earth's history.

However, like many of the Earth's natural processes, the presence of human society has changed erosion. In many cases, it's occurring much more quickly than normal. *Sheet erosion* occurs when an entire layer of topsoil is removed through the force of wind or water. This usually happens when the soil is bare, because there are no plant roots to hold it in place.

*Gully erosion* occurs when the flow of water is intense, cutting gullies that look like small canyons into the land. An important factor in understanding erosion is the effect of speed. The faster the wind or water is able to move, the greater erosive force it will have.

 **DEFINITION**

**Erosion** occurs whenever soil is moved by the forces of wind or water. **Sheet erosion** occurs when an entire layer of topsoil is removed. **Gully erosion** is caused by moving water carving small channels into soil, which gradually enlarge into ditchlike gullies.

What specific human impacts can accelerate erosion?

- Removing trees, which serve as natural windbreaks.

- Planting crops in perfectly straight rows.

- Leaving land bare and unplanted.

- Straightening the course of a river.

Remember, soil forms slowly over millions of years. If the soil is lost—blown away by the wind or swept away by moving water—it cannot be replaced.

## The Ultimate Degradation

The most extreme form of land degradation is its conversion into desert. Technically, this can occur in any kind of ecosystem, but it is most likely to happen to one that is already pretty arid, such as a grassland.

The most common cause of *desertification* is overfarming. This depletes the already fragile soil of nutrients. Raising animals can also increase the risk of desertification. The overgrazing of cattle can strip the land of vegetation faster than it's able to regrow, hastening erosion.

The obvious solution would be to provide an ample water supply through irrigation and replace the nutrients with fertilizer. While these techniques can help to slow and reduce desertification, when done improperly, they can make matters worse. Irrigated water is not pure; it contains dissolved minerals like calcium and magnesium. When land is overirrigated, these minerals build up in the soil. This process is called *salinization,* because the effects are similar to adding salt.

 **DEFINITION**

**Desertification** is the gradual conversion of an arid ecosystem into a desertlike state, including low biodiversity and dry soil. **Salinization** is caused by a buildup of salt ions in soil, such as sodium, magnesium, or calcium, usually due to rapid water evaporation.

# Save Our Soil

The biggest moral of this chapter is that soil is a precious resource, one that we need in order to grow the food to support our increasing population. Given how long it takes for soil to form from its original parent material, rock, simply waiting for the resource to replenish itself is not an option.

Like many lessons in dealing with the environment, this is one that humanity had to learn the hard way. One of the most recent examples of this occurred in the United States in the 1930s—an environmental disaster known as the Dust Bowl.

The Dust Bowl was especially tragic in that the ecosystem it occurred in, the Great Midwestern Plains of the United States, had some of the deepest and most fertile topsoil on the continent. As farming and agriculture expanded into the region, many poor decisions were made regarding the use and management of the land and oil.

- Monocultures: many farms grew only a single crop.

- No crop rotation: the same crops were grown in the same place year after year.

- Overplowing: native deep-rooting plants were removed, and the soil was loosened.

- Overplanting: nutrients were withdrawn from the soil faster than they were replaced.

- Overgrazing: cattle and sheep were allowed to eat the native plants down to the roots.

These farming practices made the soil vulnerable, but a severe multiyear drought caused the Dust Bowl. Plants wouldn't grow, the soil dried to dust, and wind carried it away in an extreme example of sheet erosion. Whatever plants were able to grow were decimated by swarms of grasshoppers desperately looking for a food source in the increasingly arid landscape.

As the country struggled to deal with the Great Depression and the Dust Bowl, many important lessons were learned. The government began making rules and recommendations for agriculture, adopting practices that would lessen the stress on the soil and hopefully avoid a future disaster.

## Shelter from Wind

Part of the New Deal of President Franklin D. Roosevelt was the Civilian Conservation Corps. Unemployed young men were targeted by this program and were put to work doing a variety of jobs related to the conservation of natural resources. One example was "Roosevelt's Tree Army." Rows of trees were planted throughout the Midwest in an effort to slow winds, hold soil in place, and reduce erosion.

On a smaller scale, trees are often planted around the edges of farms to serve as windbreaks. Remember, erosion is most likely to occur when wind and water are able to move down straight, uninterrupted paths. Trees and shrubs interrupt this flow.

Providing ground cover is also an effective way of reducing erosion. Something as simple as mulch placed over the top of plowed soil (in the bare areas between crops) greatly reduces wind erosion. Some farms will intentionally allow fields to grow wild during an off season or off year. The deep roots of weeds and wildflowers hold the soil in place until planting can occur again.

## Slowing Down Water

Reducing wind is one way of lessening the risk of erosion, but no tree line can account for the effects of running water. One additional measure farmers can take is to eliminate the traditional method of plowing and planting crops in straight lines.

Contour plowing actually follows lines of changing elevation across a sloped landscape. Have you ever looked at a contour map? In practice, a farm that uses contour plowing should have very similar patterns. This greatly slows the flow of runoff water during a rainstorm, reducing erosion. This is another measure encouraged by the government after the Dust Bowl, and one of the most successful in terms of reducing runoff.

Terracing takes this same idea a step further. Instead of simply plowing along lines of changing elevation, flat surfaces are carved into a slope, creating an effect that resembles stairs. This practice is more common in areas with steep hillsides, and is also very effective at reducing erosion from water running downhill.

*A terraced farm looks like steps have been carved into the hillside. This landscape will experience much less water erosion than a smoothly plowed hill.*

## Change Is Good

As we know, intensive farming can deplete soil of nutrients over time. All plants need these nutrients to grow, although they don't all need them in the same amounts or at the same rate. When farmers practice monoculture, growing the same crop continuously, the soil is likely to be disproportionately stripped of a few nutrients that plant is dependent on. A carefully planned crop rotation strategy can ease this demand on soil.

One common pairing seen in agriculture is to grow corn during one season, then a legume like soybeans or alfalfa the next. Legumes are plants that have a symbiotic relationship with a specific bacterium in their root systems. The bacteria help them to take in normally unusable forms of nitrogen, such as that found in the air, and convert it into a usable solid form. This is called nitrogen fixation, and is a great way to replenish nitrogen in the soil that is depleted much more quickly by non-legumes like corn.

This method is preferable in many ways to adding synthetic fertilizer, because the nitrogen is absorbed more readily into the crops and stays in the soil longer. It also saves the farmer money in the long run, as he won't need to purchase as much fertilizer.

## Organic Growing

Despite the tremendous gains in crop yields achieved during the Green Revolution, some food producers have opted against many of its prescribed inputs. Pesticides and herbicides are known to impact other species, and their residues can be found on the crops we eat. Synthetic fertilizers, while effective at providing for the short-term nutrient needs of plants, can run off and impact nearby water supplies.

 **CASE STUDY**

The Dirty Dozen is a list of produce foods that have been found to be the most contaminated with pesticide residue. The list is meant to be a guide as to which foods the consumer should make an effort to buy organic if possible. The list is published by the Environmental Working Group, ewg.org. Included on the 2012 list were mostly leafy vegetables and those in which the skin is ingested. Examples include apples, celery, sweet bell peppers, lettuce, and cucumbers. Items of produce that are lowest in pesticide residue are often those that have the skin removed, like onions.

Organic farming seeks an alternative method of growing that doesn't rely on any of these artificial measures. In place of synthetic fertilizers, organic farmers may use animal manure or carefully planned crop rotation to preserve soil nutrients. Pesticides are still used, but not synthetic ones. For example, instead of planting Bt corn, organic farmers will spray their corn with actual *Bacillus thuringiensis* cultures. In this case, the end result is the same (keeping the corn borer pest at bay), but the method is different. In place of herbicides, farmers may use cover crops, mulch, mowing, or even heat to kill invasive weeds.

The United States Department of Agriculture (USDA) has adopted a set of standards that any food advertised as organic must follow. Products that meet these requirements will be stamped with a certification seal from a third-party organization, such as Oregon Tilth.

Plant agriculture has changed a great deal in the last several decades. It has become an exact science—one that involves the input of water, fertilizer, and pesticides to achieve the maximum possible yield. This has come at a cost, however, as soil degradation continues and chemical residues are appearing on our produce.

## The Least You Need to Know

- Malnutrition occurs whenever a diet is lacking one or more essential vitamins or minerals.
- Food production has increased remarkably since the Green Revolution, primarily due to increased inputs like pesticides and irrigation.
- Soil occurs in different layers, called horizons. The layer most relevant to food production is topsoil, a mixture of organic material and minerals from weathered rock.
- The biggest cause of soil degradation is erosion. Farming techniques that slow the movement of wind and water across land will reduce this impact.
- Organically grown foods must abide by certain requirements, including not using any genetically modified organisms, synthetic fertilizers, or synthetic pesticides.

# CHAPTER 12

# Animal Agriculture

The Green Revolution ushered in a brave new world of irrigation, intensive fertilization, and pesticide application in plant agriculture. Not long afterward, animal husbandry practices began to be similarly industrialized with a common purpose—to maximize yields and profits.

Consumption of meat and dairy is a luxury most associated with the developed world. Animal products are expensive and simply not available to those under a certain income level, unless they are able to hunt wild game or fish for themselves. Additionally, producing meat is very resource intensive. A pound of meat requires about 16 pounds of grain to produce.

Why are animal products so expensive and difficult to produce? The answer actually lies in a basic ecological concept: the energy pyramid.

### In This Chapter

- The practices used to maximize livestock production
- Animal welfare concerns in factory farms
- How slaughter and euthanasia of animals are conducted
- Alternative methods of raising livestock

*As omnivores, humans can consume directly from the producer level (plant-based products) or from the primary consumer level (meat and dairy products).*

Plants are producers and are at the bottom of the ecological pyramid. They absorb energy directly from the sun and use photosynthesis to convert this energy into a chemical form, sugars. When humans or other animals directly consume plants, they are able to digest and absorb all the calories present. However, when we feed the plants to herbivores, only about 10 percent of the energy in the plants makes it into the muscle tissue and fat of the animal for us to eat later. The rest is lost as heat!

# Factory Farming with Animals

Simply put, raising animals for food is an expensive, time-consuming, and resource-intensive process. These foods, however, tend to be much more valued than plant-based foods. What is your favorite meal? Chances are the dish contains at least one animal product—cheese, eggs, or meat. With this demand, farmers have looked for ways to raise animals as quickly and cheaply as possible.

When many people think of farms that raise animals, they think of green, rolling fields dotted by cows, pigs, or sheep lazily grazing. A nearby barn may have a few workers who milk the cows or care for the newborns. Unfortunately, this idyllic image is rarely the case in modern livestock production.

Remember, with any kind of production, there are external costs and hidden costs. The changes in livestock agriculture that have emerged since the Green Revolution have had the positive effect of increasing yields and lowering prices, but there are hidden costs as well, borne by the environment and the animals themselves.

**A LOOK BACK**

The roots of the factory farm actually lie in Britain. A law passed in 1947 gave additional financial incentive to farmers who could increase their production rates. The goal was to reduce the country's dependence on imported meat. Meanwhile, in North America, farmers had realized that keeping their chickens indoors dramatically increased their survival rates. The practice grew, and eventually spread to hog and dairy cattle. Britain imported many of these ideas, which then spread throughout Western Europe.

## Milk

Two common characteristics of all mammals is the ability to give birth to live young and to produce milk. These two traits are connected; milk production doesn't begin until the animal gives birth. *Dairy cattle* are animals that have been bred and selected specifically for milk production. They're no different than other mammals, though, and must be impregnated as often as possible to promote the production of milk.

Most dairy cows are raised indoors. Enclosed sheds give milk producers the ability to control and monitor every aspect of their environment, from feeding to light levels. Cows are grouped and herded into mechanized milking stations twice a day. After one group of cows has been milked, they're herded away and another group is brought in. If the operation is large enough, it will be able to milk continually 24 hours a day. These truly are milk factories.

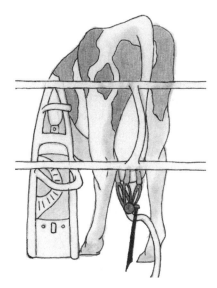

*Dairy cows are herded into a milking station, where metal bars lock them in place. A machine is attached to their udders, and they're milked mechanically.*

Once the milk is collected, it's homogenized and pasteurized. *Homogenization* is a process that mixes up the milk so the fat molecules are evenly distributed and don't separate from the other components of the milk. *Pasteurization* heats the milk to a temperature below its boiling point, killing as many of the bacteria in the milk as possible. This grants the milk an extended shelf life.

 **DEFINITION**

> Not all cows are the same. **Dairy cattle** are breeds selected specifically for the highest milk production possible. **Beef cattle** are breeds selected for their size and growth rate. When milk is collected, it undergoes **pasteurization,** a heating process that kills bacteria, and **homogenization,** mixing and processing the fat so it's evenly distributed and does not separate.

There are two significant issues faced by factory farms that produce milk: manure and disease. Cows produce a great deal of manure, and large farms will have hundreds of cows. The manure is frequently drained out and stored for use as fertilizer or for methane production. However, for at least part of the day many of the cows are walking in and around manure, resulting in diseases that can spread quickly. To overcome this problem, cows are often administered subtherapeutic antibiotics. These are added to their food regardless of whether they are ill; it's a preventative measure that has been shown to promote growth.

## Veal

The process of continually impregnating dairy cows not only results in milk, but also a lot of new calves. Female calves are kept and raised—they will become the next generation of milk producers when they are old enough. But what about the males? After all, a farm will only need a few bulls, enough to produce the sperm needed to impregnate the females. Most of the male calves will be sent to a slaughterhouse and made into veal.

Veal is a luxury item, a specific type of beef that comes only from calves. Customers have two expectations when it comes to veal: that it's tender, and it has a paler pink color than regular beef. The tenderness of the meat comes from raising the calves in enclosed hutches. They are given very little room to move around in, ensuring that the muscles are not able to strengthen and develop.

The pale color of the meat is actually a sign of anemia. The calves are given a milk replacer that is intentionally iron deficient. The lack of iron results in their bodies not being able to produce a normal amount of red blood cells, resulting in much lighter-colored muscle tissue.

## Beef

There are some important differences between the way beef and dairy cattle are raised. First, beef cattle are not bred specifically for milk production, but rather for size and growth rate. Second, beef cattle don't need to be herded into a milking station twice a day.

Most beef cattle are started on pasture or rangeland. Once they reach a certain size, they're transferred to a high-density feedlot. Raising them outdoors frees the farm from constructing and maintaining the warehouses used in milk production. This also allows beef cattle operations to be very large, and they can easily occupy hundreds of acres of open land.

The land on a feedlot is divided into fenced-off paddocks. Each paddock is surrounded by a trough filled with grain—plant products like corn, barley, and alfalfa. These feeds are very energy-dense, and result in the animals accumulating fat very quickly. This produces the marbling effect desired by consumers looking at cuts of beef. As on a dairy farm, the intense crowding of the animals necessitates the use of antibiotics to prevent disease.

## Eggs

Most livestock animals spend at least part of their lives in confinement, and none more so than layer hens, chickens bred to lay eggs. Layer hens are raised in battery cages in groups of four. Each hen gets an area about the size of a sheet of 8.5 × 11–inch paper.

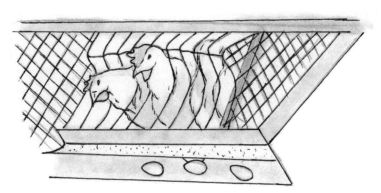

*Layer hens are often raised in cramped battery cages for their entire lives.*

The restricted space sparks an unusual amount of aggression among the hens, so to prevent injuries caused by fighting, each hen is debeaked and declawed. Light levels within the warehouses are also kept pretty low in order to minimize fighting and cannibalism.

Hens generally are kept in production until they reach about 12 months of age, when their production begins to decline. The demand for egg production is very taxing to their bodies, with one of the biggest health problems being osteoporosis, or weakened bones due to calcium loss. When a hen reaches the end of its life in egg production, it's slaughtered and used for flavorings, soups, or pet foods.

**CASE STUDY**

Of all the industrialized animal agricultural practices, battery cages have received the most criticism and scrutiny. The European Union banned their use outright as of 2012. The directive, passed in 2000, allows the use of "enriched cages." Hens have additional space to move, and a separate area to nest in. Litter and perches must also be provided. Some countries, such as Switzerland, have banned the use of cages completely.

As is the case with dairy and beef cows, only the females are used for production. When hatched, chicks are sorted into males and females. The females are kept and caged; the males are euthanized and discarded. Antibiotics are a necessity in battery-cage operations, due to the rapid spread of disease.

## Poultry

Broiler hens are special breeds that grow quickly and reach large sizes, much like beef cattle. Specifically, they're bred to have large breast and thigh areas, as these are the two most valuable cuts of meat on the bird. In fact, their bodies are so disproportionately large they are physically unable to mate, and must all be artificially inseminated!

Broiler hens reach their full weight in an astonishing two months. This is not the result of any hormone treatment, as these are completely illegal in all poultry production. Rather, it's simply a product of decades of selective breeding.

Battery cages are not used to raise broiler hens. The potential for muscle bruising from fights or rubbing up against the cage is too high. Instead, the hens are raised in enclosed warehouse pens. The amount of individual space they have is not much different than in the cages, but they are able to move around a bit more.

## Pork

Hogs are almost exclusively raised in enclosed warehouse pens. They're very sensitive to heat and sunlight, especially the pink-skinned breeds. Pigs are omnivores, so they're fed a mixture of different foods, including corn, soybeans, and meat or bone meal.

Some operations keep each hog within its own small pen, while others place them in small groups of five or more. Special pens are built for sows with piglets, called farrowing crates. These have a little extra room so the sow can lie on its side and nurse the piglets. Piglets are weaned from the mother after a few weeks and placed in their own pen.

The crowding creates issues similar to the hens: disease and fighting. Disease is handled with antibiotics. Each piglet will have its tail removed, as they will bite each other's tail under the crowded conditions.

# Unexpected Consequences

While producing animal products in industrial farms has resulted in overall higher yields and lower prices, there have been a few criticisms raised.

The first is whether the animals are humanely treated. Stress levels and pain are always difficult to measure with animals, but in the case of poultry and swine there is certainly a high level of stress, as can be seen in the violent behaviors they exhibit.

The second issue revolves around human health concerns. A tremendous amount of antibiotics are used in raising livestock—about 80 percent of the United States' total annual production. The danger arises as bacteria begin to evolve the ability to resist these antibiotics. Penicillin, the first antibiotic isolated for human use, is now shown to be ineffective in 25 percent of strep throat cases. Additionally, cattle have been shown to be more likely to carry E. coli O157:H7, the strain of bacterium that causes food poisoning—most often found in cattle that are fed high-calorie grain diets instead of grass. About 50 percent of chicken breast meat is contaminated with the same bacteria.

## Contamination

Food-borne illness, generally called food poisoning, is caused by any microscopic organism that's ingested and causes symptoms in the digestive system. Since these diseases are transmitted by ingestion, they're closely tied to the food we eat—how it's prepared, stored, and cooked.

There are several bacteria, viruses, and parasites that can cause food poisoning. The most common and prevalent are bacteria: E. coli and salmonella. When an animal contracts either of these bacteria, they work their way through the intestinal tract, eventually becoming excreted in feces. This is how these diseases spread, through fecal contact.

How does this relate to meat and dairy production? Remember, these animals are grown in crowded conditions and are continuously exposed to each other's feces. If one animal in a feedlot has E. coli, there's a good chance many of the others do as well. When the animals are brought in for slaughter and butchering, the organs are removed, including the intestines. The animals are also sprayed clean to remove any feces on their skin from the feedlot. However, this process is far from perfect. Even a tiny amount of feces that splatters into the muscle tissue will contaminate it. This is why each type of animal product has a minimum temperature to which it must be cooked in order to kill off any bacteria or other microbial contamination and make the food safe to consume.

**CASE STUDY**

Food recalls have become increasingly prevalent in the last several years. These occur most often when a strain of bacteria, like salmonella, is detected in a batch of the product. These recalls are not restricted to animal products, but have also included plants like tomatoes and spinach. How does this contamination occur? Remember, fertilizer is sprayed every spring to increase the nutrient level of the soil. If the fertilizer hasn't been sufficiently treated, it may still contain fecal bacteria, which can persist on the plants even after they're harvested.

## Manure

Dealing with the continuous production of manure by the animals is one of the greatest challenges facing a large factory farm. Many farms will temporarily store the manure in open storage ponds on the property. The manure will be removed and used as fertilizer during the spring planting.

There are several environmental issues involving manure and storage ponds. The most noticeable is the smell. The odor is incredibly strong, and a major nuisance for rural communities around these farms.

Another issue with the manure is runoff. When it's sprayed onto the fields in the spring, not all of it remains in the soil. Heavy rainfalls can cause the manure to escape into nearby streams, rivers, or lakes, where it can interfere with the normal communities present. Traces of antibiotics have also been found in this manure, and can accumulate in the soil and water sources over time.

## Slaughter

Death is an expected part of raising livestock for food. At some point, animals will reach full size (called "slaughter weight") and will need to be killed, butchered, and prepared for human consumption.

The process is pretty similar for all livestock animals. Actual death occurs when the animal's throat is cut, causing a sufficient amount of blood loss to result in brain death. The federal Humane Slaughter Act requires that animals be rendered unconscious beforehand. This is done one of two ways. The first method is to use a bolt gun, which strikes the animal with a large, heavy metal rod, inducing unconsciousness. This is used only on large animals—hogs and cattle. Animals may also be stunned by electric shock.

After death, the animals may be briefly placed into a bath of boiling hot water. This shocks the skin and muscles, enabling the feathers (or hair) to be removed from the carcass more easily. The organs are removed, and the meat is butchered and shipped away to be sold.

Kosher and halal slaughtering are based in Jewish and Islamic religious tradition. The method is similar to the previous process, but no stunning occurs before the throat is cut. The slaughter must be performed by someone well trained in the appropriate religious laws.

## Downers

Animals that are injured or become ill before they reach slaughter weight, referred to as "downers," need to be euthanized. This term literally means "good death" and is meant to be a humane way of killing an animal that's suffering from unrecoverable health issues. Ideally, any pain and distress suffered by the animal should be minimal, and death should be achieved quickly.

While there's a federal law dealing with slaughter, there is none for euthanasia. State laws actually set the standards for how animals can be euthanized. Here's an example set of acceptable techniques for killing hogs, as defined by the American Association of Swine Veterinarians.

- Suffocation in carbon dioxide

- Gunshot

- Penetrating captive bolt gun

- Electrocution

- Lethal injection, administered by a veterinarian

- Blunt trauma

# What Are the Alternatives?

In many ways, the factory farms seem like nightmarish places. Animals living in cramped, confined conditions, with disease so widespread that antibiotics must be distributed just to preserve the herd. Huge pools of manure accumulate at the facilities.

Proponents of industrialized agriculture will argue that factory farms are a necessary evil, the only way to keep up with the demands of a rapidly growing world.

But there are several alternative production methods used for each form of animal product. Each provides different enrichments to the conventional system. Some address the treatment of the animals, others what the animals are fed, and still others the overall care given to the environment.

## Enrichments

There are no universally accepted standards for animal welfare; no definition of what is completely humane or "fair." As an example, we'll look at one set, the 5-Step animal welfare rating standards from the Global Animal Partnership.

Step 1 is defined as "no crates, no cages." This provides at least enough space for the animals to move around freely. This is similar to the "cage free" designation given to some products.

Step 2 is an enriched environment. Animals are given objects to interact with, such as a bale of straw for chickens to peck at.

Step 3 provides outdoor access to animals in indoor warehouses.

Step 4 provides additional outdoor access, including the ability to roam, forage, peck, and wallow. This is similar to the "free range" label seen on some meats and eggs.

Step 5, the highest level, places the health and well-being of the animal above all economic concerns.

*Free-range cattle are given a great deal of outdoor access. A much more significant portion of their diet comes from grass, instead of corn or other grains.*

While each of these steps provides an enriched environment for the livestock, they add additional costs to the production process. Animal products made under conventional practices will be significantly less expensive than those that are free range.

## Organic

As is the case with plant agriculture, organic production doesn't utilize any synthetic chemicals. While animals aren't given fertilizer and pesticides as plants are, the organic standards still apply to their feed.

**CASE STUDY**

Whether there are any nutritional differences between organic and conventional foods has been a raging debate for years. However, with milk there are some distinct chemical differences. Conjugated linoleic acid (CLA) is often found at significantly higher levels in organic milk. This is an omega fatty acid that may be beneficial in numerous health conditions, such as heart disease and cancer. There are some differences in vitamin levels, too, especially vitamins A and E.

Organically raised animals must be given organic feed. This means grains that haven't been given any synthetic fertilizers or pesticides. Organic farming also prohibits the use of antibiotics and hormones in feed. There are also requirements for outdoor access; for example, cattle must be out on pasture for at least a third of the year.

## Polyface Farm

Polyface Farm is a family farm located in Swoope, Virginia. It is run by Joel Salatin, and is modeled after naturally occurring ecological systems.

The animals on the farm are pastured, and Joel practices a rotation system. For example, a herd of cattle will be allowed to freely graze on a specific paddock for a few days. They'll then be herded away to another area of the farm. Once the cattle have been removed, layer hens will be brought in. The hens peck through the field, naturally removing many of the insects and parasites that accompany the cattle and are left behind. This also gives the grass on the field time to regrow and recover before it's grazed again. The scratching of the hens on the field also helps to press the cow manure into the soil, decreasing runoff and encouraging natural fertilization.

The idea behind the farm is to provide the animals an environment that suits their physiological design. For example, cattle are designed to eat and digest grass; thus, they're allowed to graze as much as possible. Poultry are designed to peck and eat insects, and they are given opportunity to do so. This farm has received a great deal of acclaim for its sustainability practices.

As the Green Revolution has changed plant production, it has also left an indelible impression on animal production. However, this raises a great deal of concern, as animals, like us, have the ability to feel pain and discomfort. We're also more acutely exposed to some of the contaminants of animal factory farming, including the emergence of antibiotic-resistant bacteria and food-poisoning outbreaks.

## The Least You Need to Know

- As a result of the 10 percent rule, animal products require a great deal more time and resources to produce.
- Dairy cattle, layer hens, and broiler hens are all confined indoors for most or all of their lives. They're administered antibiotics to promote growth and prevent disease.
- Regulations regarding animal welfare, slaughter, and euthanasia are pretty sparse and tend to vary by state.
- Food poisoning is a direct result of fecal contamination of food. A frequent source is animal manure, either from slaughtering or from spreading on crops as fertilizer.
- Animals raised under organic guidelines are only fed organically produced grains. Antibiotics and hormones are not allowed.

# CHAPTER 13

# Fishing and Aquaculture

In 1497, an explorer named John Cabot sailed from England for the New World and landed on the island of Newfoundland, in what is today Canada. One of the most remarkable things about the area, according to Cabot, was the incredible amount of fish in the sea. In fact, as he recorded in one of his journals,

> The sea there is full of fish that can be taken not only with nets but with fishing-baskets.

The fish Cabot was referring to were primarily Atlantic cod. These are large fish, reaching 6 to 7 feet and weighing 200 pounds at full maturity. This account is verified by another English fishing captain, from a journal written over a century later:

> [The cod are] so thick by the shore that we hardly have been able to row a boat through them.

This goldmine began to be exploited to its fullest in 1954, when the British ship *Fairtry* docked at Newfoundland. The ship was an innovation in that it was equipped with a fish trawler, a processing plant, and even its own storage freezers.

## In This Chapter

- Methods of harvesting fish in the wild
- Problems caused by overharvesting of fish
- How fish harvesting regulations are established
- How fish are farm raised

The *Fairtry* was only the beginning. Soon fleets of trawlers from the world's superpowers began converging on the waters surrounding Newfoundland. Enormous harvests of cod were caught, peaking at over 800,000 tons in 1968. Then the harvests began to decrease, leading Canada to extend its territorial waters and evict the fleets of the other countries. They continued to drop precipitously until 1992, when all cod fishing was ceased by law.

Cod are not the only aquatic organism to face population collapse as a result of overharvesting by humans. Many other fish and shellfish have experienced similar fates. How are these fish caught at such tremendous rates? Why are some fish more costly than others to catch and process? Should we move toward farmed fish instead of wild-caught? These questions all continue to be debated as we struggle to find a way to harvest fish sustainably.

# Harvesting from the Sea

Of course, the traditional method of obtaining fish and shellfish is to catch it wild. Commercial fishing is a huge industry, bringing in millions of metric tons every year. As a reference, the total world fish biomass is estimated to be between 0.8 and 2.0 billion tons, according to a study in the journal *Science.*

Each species is caught in different amounts, is used for different foods, and demands its own price. Here is a list of the top five species (by weight) caught off the coasts of the United States.

| Species | End Product |
| --- | --- |
| Alaska pollock | Breaded fish products |
| Atlantic menhaden | Fish oil, fertilizer |
| Pacific cod | Whole fillets |
| Pink salmon | Whole fillets |
| Squid | Whole, rings (calamari) |

There are thousands of species of fish, mollusks, and crustaceans in the oceans and lakes of the world, but the commercial fishing industry relies on only a handful. These highly valued species are under constant risk of population collapse, much like the Atlantic cod. In order to understand how much pressure our demand places on these populations, you must first understand how they are caught.

## Pole Fishing

When you think of commercial fishing operations, *pole fishing* very likely doesn't come to mind. This is what people do for fun on their vacations, right? But there are commercial operations (such as tuna) that rely on pole and line fishing, and they're able to pull in pretty impressive harvests using these methods.

Pole and line fishing is typically used for fish that occupy the upper layers of the waters. Each pole has a line, at the end of which is a feathered jig (lure) and a hook. Chum—ground-up fish and other seafood—is thrown into the water as bait, whipping the tuna into a sort of feeding frenzy. Once this begins, the fishermen lining the ship's decks begin pulling tuna out at an astonishing rate.

*An experienced crew of pole-fishers working amid a feeding frenzy can reel in hundreds of fish in a very short time.*

How can this kind of fishing possibly be commercially viable? Remember, the hooks are barbless. Once the fisherman pulls the fish onto the ship, it's immediately released from the hook and herded into a storage pool. The line is immediately recast. An experienced fisherman can pull in several fish in just a few minutes!

## Trawl Fishing

*Trawling* is the act of pulling a net through an area of the ocean behind one or more boats. Trawls have been around for hundreds of years, going back to the fifteenth century. One of the biggest criticisms of the practice is its lack of selectivity. Simply put, lots of other species besides the targeted one are caught. A study of trawl fishing in Australia found that an estimated 177 tons of aquatic organisms were accidentally caught and discarded yearly.

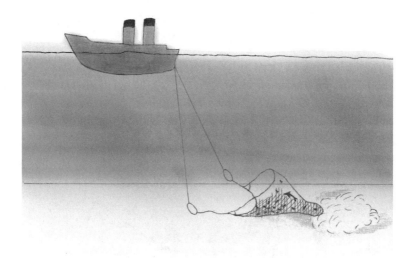

*Bottom trawlers will drag a large net through an entire layer of the ocean. The depth of the net depends on the targeted species.*

*By-catch,* the term for these unintentionally caught organisms, is somewhat controlled by altering the mesh size of the net at the end where fish are retained. This works best when the targeted fish is large, so the mesh of the net can be enlarged to allow smaller fish to escape. However, this doesn't work when the target species is small. Shrimp trawling, for example, has one of the highest rates of by-catch.

 **DEFINITION**

> Fish caught by **pole fishing** are pulled directly from the ocean by teams of fishermen using poles with hooked lures from a stationary ship. **Trawling** is when a large mesh net is dragged through the middle or bottom layer of the ocean. Any species large enough is trapped, including **by-catch**—fish that are unwanted and unintentionally caught by a trawl net.

Trawling has other environmental side effects as well. Many forms of trawling involve dragging weighted gear over the bottom of the sea. This causes a tremendous amount of damage to the habitat of the sea floor, including destroying coral and seaweed. Disturbing the sea floor in this way also stirs up a lot of sediment, creating huge plumes of sand and silt in the water. Not all trawlers have this problem. Midwater trawling occurs at a higher level of the ocean, leaving the bottom undisturbed.

# Setting Limits

Wild capture of fish and shellfish peaked sometime in the 1980s, and subsequent harvests have been either stagnant or lower. Many argue that stricter regulations need to be enforced to ensure that the existing populations of fish are not overharvested, causing a collapse similar to the one seen in Newfoundland.

Are these regulations really necessary? After all, if harvests begin to drop, a moratorium can be declared. The fish can repopulate, and harvesting can begin again. The damage caused by overharvesting, however, is often irreversible. In the case of the cod off Newfoundland, their niche in the ecosystem was actually replaced by other species. Lobster and small fish called capelin began to increase in numbers. The original population of cod will likely never return to the shores of the island.

Regulating fishing activity in the ocean is no small matter. First, the ocean is incredibly vast, and tracking the activity of each individual fishing vessel is nearly impossible. Second, the ocean has been divided into international waters and territorial waters. Territorial waters are controlled by the country they surround. International waters, on the other hand, require agreement and cooperation between countries on what laws are enforced, and how.

## Laws in North America

In the United States, commercial fishing restrictions are divided into state and federal laws. Federal restrictions are administered by the National Oceanic and Atmospheric Administration (NOAA).

There are six regions of coastline in the United States, and the actual restrictions vary for each species and region. In 2012, the United States became the first country to impose catch limits for every species of fish it manages. Counting every fish in the ocean is impossible, so how are these limits established?

Every few years, NOAA conducts stock assessments of commercial species. This includes doing its own random trawls and interviewing commercial fishing operators about their catch rates for the past few seasons. This data is never exact, and disputes are commonplace.

For example, NOAA estimated that the stocks of cod in the Gulf of Maine had declined by two-thirds over a four-year time span. The agency was considering setting catch limits for this fish much lower than before, leading to opposition from local legislators and fishermen.

While these restrictions are never popular, they help to ensure that no one species is lost. One example of a success story is the summer flounder, a mid-Atlantic species that was deemed 88 percent below healthy levels in 1989. A more recent survey has shown that the species has been restored, with a population much closer to its normal levels.

## International Waters: The Law of the Sea

A 1982 United Nations convention gave all countries the freedom to fish without regulation in international waters. These were defined as areas that fell outside the 200-mile "exclusive economic zones" past the physical border of each country.

Fishing in international waters quickly became a textbook example of the tragedy of the commons. You might remember the essay from Chapter 1 of this book describing how individuals are more likely to overuse a resource in the short term if it is one they do not own. International waters are truly a common, and nations would race against each other to harvest as much from those areas of the ocean as possible.

**A LOOK BACK**

The cod wars occurred through the 1970s as Iceland declared an exclusive economic zone extending beyond its territorial waters. To enforce this zone, the Icelandic coast guard was deployed, ordered to cut the nets of any British trawlers in the area. The confrontations came to a head when the British Royal Navy began sending warships to protect their trawlers. These confrontations spurred the eventual passing of the United Nations Law of the Sea Convention.

This led to conflict during the 1990s, as most stocks of commercially valuable species were beginning to run low. Coastal countries felt their efforts to restrict fishing and conserve stocks were being undermined by unlimited fishing in the high seas.

The United Nations Law of the Sea Convention went into effect in 1994. This agreement required all member countries to take conservation measures to protect the living resources of the ocean. Many of the restrictions agreed upon in this law centered around the use of large drift nets. Between the overall decline in fish stocks and the reluctant implementation of conservation laws, fish harvests worldwide have either declined or plateaued.

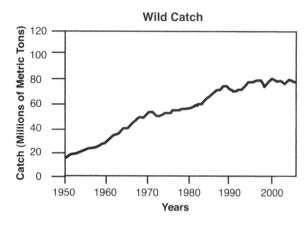

*Fish harvest peaked in the 1980s. With the collapse of many commercial fish populations, harvests have been declining ever since.*

## The Seafood Watch Program

The issues and regulations that governments and fishermen wrestle with are nearly impossible for the average consumer to fully grasp. With that in mind, a few organizations have begun issuing guidelines and suggestions for concerned consumers purchasing seafood.

One example is the Monterey Bay Aquarium Seafood Watch. Ratings are given to each species of fish and shellfish based on how they are caught and the current well-being of the population. The ratings, taken from www.montereybayaquarium.org, include the following:

**Best Choices:** Species that are in abundance or caught in sustainable ways.

**Good Alternatives:** These species are considered acceptable options, but there are concerns about either how they're raised or the health of the wild population.

**Avoid:** Species that shouldn't be purchased due to concerns about their sustainability. The environment and other species may be adversely affected by harvesting this species, as well.

The most popular seafood in the United States is shrimp. Many of these are wild-caught in the Gulf of Mexico and South Atlantic. Shrimp are short-lived species that reproduce quickly, and have little risk of population collapse. However, since they are so small, the trawl nets used to catch them also have a high rate of by-catch. They are listed as a "Good Alternative."

Canned tuna is the second most popular product among U.S. consumers. Canned tuna can actually consist of several species, such as bigeye, yellowfin, or skipjack. Tuna reproduce quickly, but the demand for them is so high that many populations are in decline. Most of this tuna is caught in trawl nets that have high rates of by-catch. With this in mind, most canned tuna is labeled as "Avoid." Exceptions are made for canned tuna that is pole-caught, since the by-catch rate is much lower.

**CASE STUDY**

One of the greatest species of concern in trawling by-catch is endangered sharks. Sharks are the physically largest group of fish and can easily become trapped in trawl nets intended for other species. Sharks are also accidentally caught and reeled in by pole-fishers. A promising new technology incorporates rare earth metals such as samarium into hooks that actually repels the sharks. Sharks are magnetically sensitive, and these metals make some of the strongest magnets known to man.

The third most popular species is salmon. Most of the salmon sold in U.S. grocery stores is farmed, not wild-caught. With this in mind, it has received the "Best Choice" rating. Seafood farming, called aquaculture, is a rapidly growing industry and is quickly becoming a favored alternative among environmentalists.

# Aquaculture: Fish Farming

As wild seafood harvests have declined or plateaued over the past two decades, aquaculture has begun to replace what the sea cannot provide. Aquaculture is the farming of aquatic organisms, such as fish, crustaceans, and mollusks.

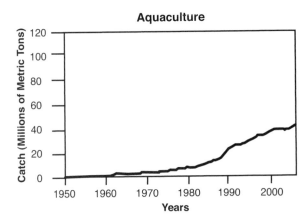

*As wild-caught fishing operations have become less commercially viable, aquaculture operations have grown in their place.*

Aquaculture can apply to both freshwater and marine organisms, but most of the industry revolves around a few saltwater species. Some species, such as abalone (sea snails), are almost exclusively farmed as the wild population has become so sparse.

## Farming in the Ocean

Mariculture is the raising of saltwater organisms. This can take place either in the open ocean, in an enclosed section of the ocean, or in large tanks.

If the farming takes place in the ocean, usually an area is netted off to prevent the fish from escaping. This is called cage-culturing. An initial stock of immature individuals or eggs will be placed into the enclosure and raised.

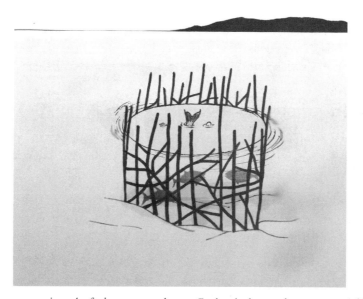

*A fish cage, or pen, is made of a long, rectangular net. Food and other supplements are sprinkled into the top of the pen, and wastes are allowed to drop out the bottom to the ocean floor.*

Similar inputs seen in industrialized agriculture of livestock apply here as well. The fish are given feed, which, depending on the species, may be either plant or animal matter. Antibiotics are applied to keep the spread of disease down, since the population is so densely packed within the enclosure. Herbicides may also be added to prevent overgrowth of algae.

**CASE STUDY**

The two most produced species in aquaculture are salmon and carp. Research in the genetic modification of these species is underway. Salmon, for example, have been genetically modified to produce more growth hormones and mature faster.

The diet of the salmon is designed to be as close as possible to that of wild salmon. Wild salmon eat krill, which gives their flesh a deep red color. Farmed salmon are fed a replacement pigment, although the resulting color of the meat is noticeably lighter.

Mariculture operations can have huge impacts on the surrounding ecosystem. Often, the fish raised in the enclosure is a non-native species. Nets are not permanent structures; they do get damaged, and the fish do occasionally escape. When this happens, there's the potential for an invasive species reproducing and taking over an ecosystem.

The single biggest environmental impact of cage-culturing is the high amount of waste produced. This waste, which also includes uneaten food, sinks to the bottom and can actually smother any organisms living in the benthic layer underneath. The waste commonly contains disease and parasites, which can also spread to native species.

## Ecology-Based Aquaculture

Aquaculture has a lot of benefits: it lowers the price of fish and seafood and reduces pressure on wild stocks. The biggest drawback seems to be the waste production. Integrated Multi-Trophic Aquaculture (IMTA) is a design that attempts to address this problem.

Instead of simply raising one species, IMTA systems raise two or more. The waste produced by one species is used as an input to help raise another. The system is actually set up like the trophic levels in an ecosystem.

Carnivorous fish or shrimp usually occupy the higher trophic levels of IMTA. Their waste products include ammonia and phosphorous. The detritus, which includes uneaten food scraps and waste, passes down to the next level. Organisms that occupy middle trophic levels like clams and oysters will use some of the detritus from the fish. Finally, the waste products make excellent fertilizer for seaweed that occupies the bottom producer trophic level. The combination of each of these three levels helps to alleviate and recycle the waste produced by the farm and minimize the impact on the surrounding ecosystem.

All fish used to be wild-caught, but as wild fish populations have collapsed, more and more are farm-raised. Attempts to regulate fishing of territorial waters have been largely too late. While aquaculture has its own set of environmental consequences, it does meet the worldwide demand for fish without permanently destroying existing wild stocks.

## The Least You Need to Know

- Into the 1980s, the vast majority of consumed fish were wild-caught, many through bottom-trawling. This practice, in addition to damaging the ocean floor, has a high rate of by-catch.

- Each country has territorial waters that it establishes catch limits for. International waters fall outside these areas.

- Aquaculture, the raising of fish in ponds or in enclosed pens in the ocean, is rapidly becoming the preferred method of producing seafood.

- Integrated Multitrophic Aquaculture involves the farming of multiple species (at different trophic levels) at once to minimize the pollution produced by the waste of a single species.

# A Look at Energy

Part 4 covers energy. Production of electricity and fuels might be the single largest cause of environmental degradation. In these chapters, I focus on the energy source we rely upon the most: fossil fuels. We also examine how coal, oil, and natural gas are extracted, processed, transported, and eventually used by the consumer.

Fossil fuels are a resource we won't have forever, and the search for alternative sources of energy begins with nuclear power. Nuclear energy initially created a great deal of optimism following the two world wars, but a series of major accidents has brought the danger of this fuel into the light.

In the long run, our fate may rely on our ability to harness renewable sources of energy: solar, wind, hydrokinetic, geothermal, and more. Each of these sources has some drawback preventing it from becoming one we rely upon more, but they all play an important part in developing a long-term energy source that's not as environmentally degrading as the fossil fuels.

# Fossil Fuels

Throughout human history, our society has utilized many different forms of energy. In the early ages, humans relied primarily on muscle power—manual labor by themselves, servants, or animals. Heat was primarily generated by burning wood, although some civilizations had begun using coal very early on.

Beginning with the Industrial Revolution, the heat generated by burning coal was harnessed in a different way. Engines were invented that could take steam produced by boiling water and use it to pump pistons. In this way, entire trains could be driven by shoveling coal into a steam engine.

In the time that's passed since the Industrial Revolution, our society has gotten much better at extracting and utilizing the potential energy locked in coal, crude oil, and natural gas for a wide variety of applications—electricity, heat, and transportation. This advancement has also made us very dependent on a nonrenewable source of energy, and one that produces a great deal of air and water pollution.

## In This Chapter

- The basic chemical composition of coal, crude oil, and natural gas
- Geologic and biologic processes that formed each of the fossil fuels
- How fossil fuels are extracted from the Earth
- Why each fossil fuel is vital for our current society
- The relative environmental damage caused by extraction

# Coal

The word *coal* has its roots in Old English, from a term that means "mineral of fossilized carbon." While this is a good start, there's much more embedded in this mineral than carbon. In order to understand exactly what coal is made of, a little background in how it's formed is necessary.

 **A LOOK BACK**

Coal was first widely used in China. Europeans were rather astonished at its ability to produce heat, as shown by the writings of Marco Polo: "Throughout the whole province of Cathay are a kind of black stones cut from the mountains in veins, which burn like logs. They maintain the fire better than wood. If you put them on in the evening they will preserve it the whole night, and it will be found burning in the morning."

The roots of coal deposits actually go back hundreds of millions of years. While several geologic ages have contributed to the formation of coal deposits, we're going to focus on the Carboniferous Era. Two important conditions were present during this time (300 million years ago) that favored the formation of coal. The first is the evolution of a complex carbohydrate called lignin. Bark and the wood of trees are partially made of this substance. Animals and bacteria had not yet evolved the ability to digest this lignin, so it began to pile up as trees grew and died. The decomposer part of the carbon cycle was somewhat inhibited, so the atmosphere became richer in oxygen, increasing wildfire activity.

The second major factor during the Carboniferous period was the presence of lower sea levels. This allowed the formation of low-altitude forests and swamps throughout North America and Europe. Eventually these forests and swamps flooded and then were buried under soil. Decomposition was inhibited, so the carbon-rich plant material was slowly compressed and heated. Depending on how long these geologic conditions continued, the plant material became peat (such as what is found in bogs), coal, graphite, or even diamonds.

With this in mind, the composition of coal matches its plant precursors. The primary element in coal is carbon, but the other elements of life, including nitrogen, oxygen, hydrogen, and sulfur, are also found within.

## Where Is Coal Found?

Given its history, coal can be expected to best form in places that had low elevation and rich plant life (especially forests) during past geologic ages like the Carboniferous Era. During this era, the supercontinents of Pangaea and Gondwana were only beginning to separate.

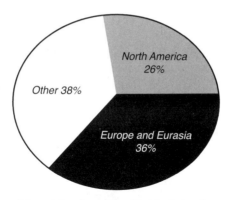

*The majority of the world's coal deposits are found in North America, Europe, and Asia. The remainder is spread throughout the continents of Australia, South America, and Africa.*

The relative abundance of coal in countries like the United States, Russia, and China has played a huge role in how these countries have developed and industrialized into the modern superpowers.

Coal is a nonrenewable resource. Over time, we could theoretically exhaust the accessible supply. How much is left in the Earth's crust? That's an extremely difficult question to answer, but it's speculated that the world may reach the peak of coal production in 2025, after which production in subsequent years will gradually diminish. For the time being, however, it's the cheapest and most plentiful of the fossil fuels.

## Digging Underground

Deep underground coal mining accounts for about 60 percent of the world's coal production. The actual process of developing and using an underground mine is pretty complex.

Once the mine is opened and accessed, a large machine digs and scrapes coal from the seams of the walls. The loosened coal falls onto a conveyor belt that returns it to the surface. The actual "mining" is done primarily by machine. The workers are there to maintain the machines and shovel the loosened coal onto the conveyor belts.

One of the biggest dangers with underground mining is explosion and collapse. Due to the work of the machine and workers, the air is continuously filled with coal dust, which is flammable. Methane gas deposits are also commonly found in the coal seams, adding to the danger. To help control this, workers will spread a nonflammable rock dust to help dampen the coal dust. Ventilation shafts are also built into the mine to help provide some fresh air.

**CASE STUDY**

In April 2010, an explosion occurred at the Upper Big Branch Mine in West Virginia. This was considered the worst mine accident in 30 years, with 29 casualties. The investigation concluded that the mine did not meet safety requirements, including a sufficiently maintained ventilation system to remove any buildup of methane gas and coal dust, which caused the explosion. Additionally, the Mine Safety and Health Administration was faulted for not acting on safety violations that had been identified years before. Over $200 million in fines and settlements was paid as a result of this accident.

Black Lung Disease is also a fairly common side effect of working in an underground mine for a long period of time. When continuously breathed in, coal dust can accumulate in the lungs. The dust cannot be removed from the body, and causes a buildup of scar tissue in the lungs. Scar tissue doesn't function in absorbing oxygen, so the lungs become much less efficient. Most modern mines will provide workers with free respirators, but they'll often refuse to use them as it increases the difficulty of breathing in a very labor-intensive job.

## Scraping the Surface

Not all coal deposits are located this deep underground. Others are much closer to the surface and can be extracted without actually digging a deep mine. There are multiple techniques for extracting coal from surface layers.

Strip mining involves large machines removing entire layers of soil and rock to access the coal underneath. These machines are huge, reaching lengths of 80 meters (262 feet) and able to move thousands of cubic meters of soil and rock per hour.

Mountaintop removal uses explosives to blast the upper soil and rock layers from the top of a mountain. This is most commonly used in the Appalachian Mountains, such as in West Virginia. This causes a dramatic change to the land, as hundreds of feet of soil may be removed to access the coal. The entire topography of the mountains becomes lower and flatter.

## Casualties of Mining

Coal mining, whether extracted from deep underground or stripped from the surface, is very damaging to the environment. When a mine is dug, water is able to flow through it much more freely and quickly than normal. As it does so, it dissolves metals and sediments from the walls and floor of the mine. The most common of these materials is a mineral called pyrite, which is made of iron and sulfur. The metals themselves are very damaging to living organisms. Many of them are neurotoxins and are not easily excreted from the body.

The sulfur, when dissolved in water, becomes sulfuric acid. Sulfuric acid is much stronger than normal rainwater, and when it exits the mine it can quickly drop the pH of a river or body of water. pH is a growth-limiting factor for many aquatic organisms.

Surface mining can actually have some of the same effects. As the topsoil is stripped away, the underlying rock (and coal) is exposed to the elements, including rain. This not only has the effect of increasing erosion, but the material that erodes has many of the same contaminants as acid mine drainage. Surface mining has the additional side effect of completely stripping the vegetation from the land. What was once a lush forest or a green mountain summit is turned into a barren moonscape of rock and clay. A law passed in 1977, the Surface Mining Control and Remediation Act, implemented a tax on all coal extracted through surface mining that helps to pay for remediation efforts.

## Why We Need Coal

Coal was initially valued as a compact fuel that could produce a lot of heat. It was commonly used by blacksmiths to melt and forge metal. Some homes even used it as a primary source of heat in the winter months. As the Industrial Revolution progressed, coal was the fuel of choice for running just about everything with an engine.

Today, coal is primarily used as a source of generating electricity. A coal-fired power plant works by burning tons of coal in a large furnace, or combustion chamber. Pipelines containing water run through the furnace, absorbing the heat. The water boils and the steam rises. The force of the rising steam is used (just like with the steam engines) to spin a turbine.

Turbines are huge, complex machines, but at their essence they are wheels that spin as steam passes through them. The turbine is connected to a generator by a metal shaft. The generator is able to take the mechanical force produced by the turbine and convert it into electric current, which can then be transported miles away and used by the consumer.

What makes coal so attractive as a source of electricity compared to other fossil fuels? For countries in North America, Asia, and Europe, it's simply a question of supply and price. Coal is cheap and plentiful in these regions. As a solid fuel, it isn't really practical for use in modern engines, so electricity is a natural fit.

Electricity rates are measured in a unit called the kilowatt-hour (kWh). The average rate in the United States is about 12 cents per kWh. As an example, consider a medium-size window unit air conditioner. The unit runs on 1,000 watts, so one hour of use would be 1 kWh and would cost 12 cents (not counting taxes and fees). Countries with greater access to coal or another fossil fuel will have lower rates, and vice-versa. Dubai, an oil-rich country, only pays 8 cents per kWh.

# Crude Oil

Crude oil bears many similarities to coal, especially in its composition and formation. But instead of being formed from the accumulation of plant matter from forests of bark-bearing trees, the source of crude oil is actually algae and plankton.

During past geologic ages, areas that contained warm, nutrient-rich bodies of water produced a tremendous amount of these two organisms. The amount of production was so great that organic matter dropped to the sea floor faster than the decomposers could recycle it. As the continental plates of the Earth shifted, many of these sea beds became buried by mud and rock. The familiar forces of heat and pressure over hundreds of millions of years converted this dense organic matter into thick, black crude oil.

The elemental composition of crude oil is similar to coal, including all the elements of life: carbon, hydrogen, oxygen, nitrogen, and sulfur. The key difference is in the types of molecules produced, called hydrocarbons. Hydrocarbons are liquid at room temperature, thus giving a whole new range of possible applications for the components of crude oil.

## Where Is Crude Oil Found?

Most of the world's oil reserves are found in North America, the Middle East, and Russia. The distribution of crude oil matches the location of warm, nutrient-rich, and biologically productive ocean ecosystems during past geologic ages, particularly in an area called the Tethys Sea.

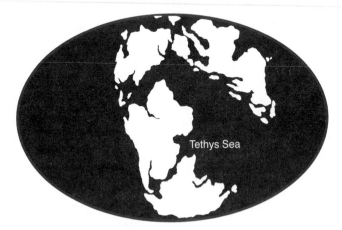

*The Tethys Sea, an ocean that existed between the supercontinents of Gondwana and Laurasia during the Mesozoic Era.*

The Tethys Sea had a series of currents that allowed a continuous upwelling of water from the ocean floor. This allowed nutrients to be continuously returned to the upper (euphotic) zone of the ocean. A combination of the plentiful nutrients and sunlight created an optimal condition for photosynthetic algae and plankton.

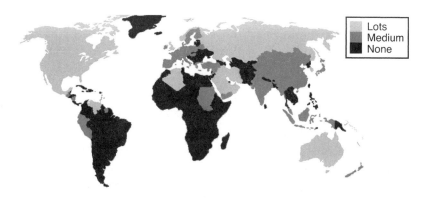

*Areas formerly within the Tethys Sea, along with the Gulf of Mexico and a few other areas around the world, have become the oil hotspots.*

Like coal, oil is a nonrenewable resource. A finite amount exists in the Earth's crust for us to extract. Estimating how much is left is difficult, as not every country is forthcoming with their geologic data. What is known is that the United States hit peak oil back in the early 1970s, and production has declined ever since. The world production is likely to face the same pattern sometime over the coming decade.

## Drilling for Deposits

The process of extracting oil is a bit different than coal, since it's a liquid. Oil can be pumped directly out of a deposit, eliminating the need for any surface or underground mining.

The process begins with drilling a large hole deep into the Earth to reach the deposit. Once drilling is complete, the hole is lined with steel pipe and cement to preserve the strength and integrity of the hole. Once the well is completed, the top is outfitted with a series of pipes and valves that collect and control the flow of oil.

Drilling can take place both on land and in the ocean. Deep-water drilling rigs contain all the necessary equipment for boring the hole, lining it, and extracting the oil from the ocean floor. Once the oil is extracted, it's temporarily stored, then pumped into a tanker truck or ship that takes it to be refined into the end products consumers use.

## The Catastrophe of a Spill

On the surface, it would seem that oil drilling should be much less environmentally damaging than coal mining. After all, the bore hole is much smaller than the entrance of a mine, eliminating the danger of acid mine drainage. Workers don't actually enter the bore hole, so the risk of death due to collapse doesn't apply.

The problem with oil, environmentally, is that it's a liquid, and as a liquid, it has a tendency to leak, spill, and escape its container. Oil spills are an almost inevitable consequence of drilling, and there have been some incredibly devastating oil spills throughout our history of using this resource.

As examples, let's look at two of the worst spills ever to occur. The first was the result of a supertanker running aground. The *Exxon Valdez* was loaded with 1.5 million barrels of crude oil, heading south toward the west coast of the United States. A combination of factors, including the presence of sea ice, an intoxicated captain, and inadequate monitoring by the Coast Guard, led to the tanker straying past the normal shipping lanes into a shallow reef. The tanker ran aground, and the rocky reef tore open its hull. Oil began leaking out of the ship, and continued to do so for several days, spreading throughout the Prince William Sound area of Alaska.

The amount of oil spilled from the *Exxon Valdez,* about 250,000 barrels, pales in comparison to the Deepwater Horizon spill in the Gulf of Mexico. This was an undersea drilling rig that experienced an explosion. Methane and other combustible gases are commonly found in oil wells. A surge of methane ignited and exploded, ripping open the pipeline into the undersea deposit. An estimated 4.9 million barrels of oil flowed freely into the ocean for nearly three months before the well was successfully sealed.

Oil has devastating effects on aquatic ecosystems. The fumes alone are enough to overcome many animals. Oil penetrates through the feathers of birds and fur of mammals, reducing their ability to insulate themselves from temperature changes and float in the water. In areas of high concentrations, it can wipe out entire populations of plankton and krill, organisms that form the basis of their food chains. Some of the oil accumulates on the sea floor, suffocating all the benthic organisms.

## Cleaning Up the Mess

Oil spills are very difficult to manage and clean up. As a liquid, oil has a tendency to spread throughout water and soil. With this in mind, much of the initial response to a spill centers on containment. In the case of the *Exxon Valdez* and other tanker accidents, ships are brought to siphon the remaining oil off the vessel. *Booms,* large floating barriers, are used to keep the oil on the surface of the water from spreading further.

The options for dealing with oil that has already escaped the immediate area are limited. Surface oil can be burned off in a *controlled burn,* which successfully removes it from the water but does produce a great deal of air pollution. Chemicals called *dispersants* work like soap by breaking up oil slicks and causing them to spread throughout the water column. These do not actually remove the oil, though, and the diluted product can still be toxic to fish and coral. Machines called *skimmers* can remove oil floating on the surface.

 **DEFINITION**

> Tools for cleaning oil spills include a **boom,** a temporary floating barrier that's laid down to prevent surface oil from spreading. A **controlled burn** sets surface oil ablaze, removing it from the water. **Dispersants** break up oil into smaller droplets, which can disperse much more easily throughout the water column. **Skimmers** mechanically separate water from oil along the surface of the water.

In many cases, much of the oil is left behind in the water. Fortunately, there are microorganisms that are able to consume and digest crude oil. Over time, it will be naturally removed from the ecosystem. The problem is, this process can be slow. If the water is warm and rich in oxygen, bacteria may be able to remediate the oil within a matter of weeks. However, if the water is cold (as in Alaska) or low in oxygen (as in the bottom of the Gulf of Mexico), remediation will take years.

The ecosystems most impacted by oil spills may well be the coastal ones. A lot of the oil washes up on shore, as a thick, gooey mass called "mousse," as the texture is similar. Coastal ecosystems are very fragile ones, with food webs built upon species that are very sensitive to pollution. Oil spills can quickly wipe out these entire food webs, leaving the ecosystem irreparably damaged for decades.

## Why We Need Oil

In spite of the inherent dangers and damage caused by drilling and extracting oil, it's an incredibly important source of energy for our society. Its value comes from the fact that we can refine and separate crude oil into many different products.

Raw crude oil itself isn't particularly useful. It can be burned for heat or electricity, but its real value emerges when it's refined into its individual components. Refineries work by basically separating these components by weight and density. Examples of petroleum products range from gasoline to paraffin wax.

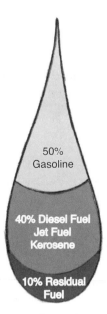

*Crude oil is a mixture of multiple components that can be separated and used for different purposes.*

Each of these individual products has different applications and advantages. Gasoline and diesel are liquid fuels that are used in just about every form of transportation. Paraffin wax is a solid material that can be easily melted and formed into a wide variety of products—everything from crayons to candles to beauty products.

This is the true value of crude oil, and our society has become incredibly dependent on this resource. Vehicles that run on fuels other than gasoline or diesel have only recently become available. Examples are cars powered by electricity and hydrogen fuel cells. Nearly every form of plastic is also made from crude oil products. How many things do you own that are at least partially made of plastic?

The versatility of crude oil and its products has caused it to become more and more in demand as the world has continued to industrialize. Gasoline, which cost less than two dollars per gallon in the late 1990s, currently costs between three and four dollars throughout most of the United States. Why? High-population countries like China and India are beginning to adopt some aspects of the American lifestyle, including car ownership. Demand for gasoline will only continue to rise, and with a finite supply, the price will inevitably follow.

# Natural Gas

The last fossil fuel is the simplest, at least in terms of chemical composition. Methane is a flammable gas with the formula $CH_4$.

The production of natural gas is a little different than coal and oil. While it begins with organic material, just like the other fossil fuels, it's produced mainly as a result of decomposition. Many bacteria produce methane as a waste product of their metabolism. In an ecosystem with a lot of accumulated plant matter, like a swamp, the bacteria are able to grow and produce a great deal of methane gas.

Swamps can be very deep, and the bacteria that live in them prefer it that way, away from oxygen. As a result, the methane doesn't always bubble up and into the atmosphere like you might expect; rather, it can be trapped and accumulate. Over time, as the Earth's plates shift and move, the swamp can be buried by layers of soil and rock. The methane gas remains behind as a deposit.

## Where Natural Gas Is Found

Due to the ability of bacteria to produce methane in a relatively short amount of time, this resource can be classified as both a renewable and nonrenewable resource. The underground sources are nonrenewable and finite. However, more can be produced above ground by harnessing these bacteria. We'll examine this process more closely in Chapter 16, on renewable energy.

Natural gas is often found in the same places as crude oil. In fact, it's often extracted (sometimes accidentally) with the oil. Remember, a surge of methane gas was behind the Deepwater Horizon oil rig explosion that caused the great oil spill in the Gulf of Mexico.

New sources of natural gas have been discovered in the last decade. Deep below the Earth's crust, under layers of a type of rock called shale, huge deposits of natural gas lie trapped. These deposits have greatly increased the natural gas production in countries such as the United States. According to the Energy Information Administration, in 2000, shale gas accounted for only 1 percent of natural gas production in the United States. Ten years later, that number is about 20 percent and rising.

## Drilling and Fracking

Natural gas can be extracted by drilling just like oil. Since it's a gas, it will naturally rise to the surface when an opening to the deposit is made. The gas is collected and directed into a pipeline, which eventually delivers it to a storage tank or directly to the consumer.

Extracting shale gas is done through a much more difficult process called hydraulic fracturing, or "fracking." The problem with shale gas is that it exists within a deep layer of rock that is completely impermeable. Releasing the gas is not as simple as drilling a hole.

Fracking works by pumping a fluid mixture into a drilled hole at a great enough pressure to cause the shale to crack or fracture. The natural gas is then able to escape the rock formation and be collected.

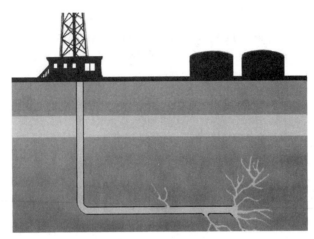

*Hydraulic fracturing drills down into the deep shale bed, then sideways to provide a well to inject fluid, creating the cracks needed to release natural gas.*

One of the biggest concerns with hydraulic fracturing is the possibility of introducing the chemicals in fracturing fluid or the methane gas itself into groundwater. The shale bed exists below the water table, so if either of these substances were to migrate upward, contamination would occur.

 **CASE STUDY**

In 2009, 13 water wells in Dimock, Pennsylvania, were found contaminated with methane. One of them exploded. Although the company that carried out the hydraulic fracturing in the area denied responsibility, they were required to provide alternative sources of water and install filtration systems. A moratorium on drilling was enacted in 2010. Was the methane really from the fracturing? The EPA collected and analyzed methane from the contaminated wells, and found it had the same "signature" of isotopes as the shale gas. One possibility is that the casings installed after drilling failed, allowing the gas to leak out at the level of the water table.

## Why We Need Natural Gas

For a long time, natural gas wasn't heavily utilized as a source of energy. Its physical state of matter, a gas, makes it much more difficult to store and transport than solid coal or liquid oil products. It wasn't until the interstate natural gas pipeline system was built, along with smaller pipelines to deliver the gas to individual homes, that its use become commonplace. When natural gas is delivered directly to homes and businesses, it's most commonly used as a source of heat in furnaces, clothes dryers, hot water tanks, and stoves.

Natural gas, like the other fossil fuels, also can be burned as a way to generate electricity. The power plants would operate much like the coal-fired ones, only with a different fuel. This hasn't been a common practice in the United States (the cost was too high), but that's changing as shale gas has driven up supply and driven down prices.

Our society is extremely energy-dependent. The cheapest and most readily available sources of energy are fossil fuels. The damage caused by the extraction and transportation of these fuels will continue until alternatives are found.

## The Least You Need to Know

- Surface coal mining is incredibly damaging to the environment, and underground mining is very dangerous to the human workers as well as to the environment.
- Oil spills are a regular problem in extracting and transporting the resource. No cleanup method completely removes the oil from water once spilled.
- Natural gas is the cleanest burning of the fossil fuels, but it requires deep drilling that may endanger groundwater supplies.
- Coal is primarily used to generate electricity, while oil produces the fuels needed for transportation. Natural gas is usually burned for household heat.

# CHAPTER 15

# Nuclear Energy

In 1886, a scientist by the name of Henri Becquerel was studying materials that glow in the dark after being exposed to light. He conducted an experiment in which a photographic plate was wrapped in black paper and then exposed to different phosphorescent materials. He believed the glow produced by these substances might be related to x-rays, and should penetrate through the black paper.

The only substance that produced a positive result in this experiment was uranium. Other scientists realized that the uranium was actually decaying into another element, releasing energy and particles in the process.

The damaging effects of radiation initially weren't known. As a result, radioactive elements like radium were given many consumer applications. For example, radium was used to paint watches, clocks, and instrument dials, because it would make them glow in the dark. Radium was even used in toothpaste and food, as it was believed to have curative powers. Unfortunately, it was discovered that the energy and particles released by these elements had the ability to penetrate living tissue and damage the DNA code within.

## In This Chapter

- The events leading up to the development of the nuclear reactor
- Environmental consequences of nuclear bomb testing
- The design of a pressurized water nuclear reactor
- The four major nuclear reactor accidents
- What nuclear waste is, and how we are dealing with it

An example of this is the "Radium Girls," a group of five women who had jobs painting the radium-based paint on watches and clocks. The radium entered their bodies, where it was treated as calcium and deposited in their bones. The radioactivity subsequently caused the destruction of their marrow and the mutation of their bone cells. No protective gear had been provided to the women.

# Nuclear Weapons

Not long after these initial discoveries were made, World War I broke out across Europe, followed two decades later by World War II. As a result, most of the research on radioactive elements centered on their use as weapons.

The explosive power of these weapons was destined to be greater than any other weapon that had ever been developed. A method of measuring these explosions was needed, and the TNT equivalent was born. A hundred tons of TNT (0.1 kilotons) were exploded in Alamogordo, New Mexico, in 1945 to serve as a sort of measuring stick for the weapons to come.

The first detonation of a nuclear device was codenamed "Trinity." The bomb was exploded in the New Mexico desert, producing a yield of 20 kilotons of TNT. Residual radiation remains on the site today, 10 times stronger than what would normally be expected.

**A LOOK BACK**

As he watched the first nuclear test, research director J. Robert Oppenheimer stated that he was reminded of a line from Hindu scripture: "Now I am become Death, the destroyer of worlds."

Everyone who witnessed that explosion was acutely aware of the awesome destructive power of their own creation. The very nature of war over the proceeding century would be completely changed by this technology.

## The Fission Bomb

The first nuclear weapons used uranium as a source of energy. Uranium is an element found toward the very bottom of the periodic table. It has a large nucleus, composed of 143 neutrons and 92 protons.

When uranium is struck by a neutron, it temporarily forms a new isotope, then promptly splits into two smaller nuclei. This reaction, called nuclear fission, releases a tremendous amount of energy.

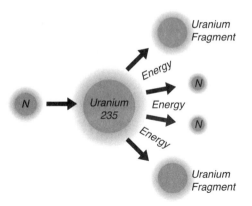

*When a uranium nucleus is split, it releases neutrons in addition to energy. If those neutrons hit another uranium nucleus, it will also split, releasing still more energy and neutrons.*

Given the right amount of nuclear material, a fission reaction should be self-sustaining. In other words, it will be a chain reaction that will continue on its own. The amount of material needed to sustain this chain reaction is called the critical mass.

Uranium ore alone is not enough to create this critical mass. The ore itself contains only about 0.7 percent of Uranium-235, the radioactive isotope. This fraction must be increased to 90 percent in order to sustain a chain reaction in a nuclear weapon. This is known as highly enriched (weapons grade) uranium.

## Hiroshima and Nagasaki

Trinity was the first detonated nuclear device. The first nuclear weapons detonated in actual combat were the "Little Boy" and "Fat Man" bombs dropped on Hiroshima and Nagasaki, Japan, in August 1945.

The two bombs had different designs: Little Boy was a uranium bomb, and Fat Man was a plutonium bomb. Their yields were 16 kilotons and 21 kilotons, respectively.

Casualties at Hiroshima alone were between 90,000 and 140,000, if later deaths from injury and radiation exposure are included. Nearly 70 percent of the city's buildings were destroyed. Six days after Fat Man was dropped, Japan surrendered and World War II was over.

## Operation Crossroads

Even though World War II was over, research and development of new and more powerful nuclear weapons did not end. Quite the opposite, as the United States and Russia embarked on an arms race to develop the most advanced nuclear arsenal as quickly as possible.

As part of this arms race, the United States conducted a series of additional nuclear detonations. Testing in the New Mexico desert was no longer an option, as the public became increasingly concerned about the risk and fallout from these explosions. An alternative site was chosen: the Bikini Atoll.

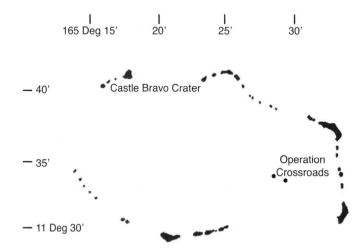

*The Bikini Atoll is part of the Marshall Islands, a small Pacific chain between Hawaii and Japan.*

"Operation Crossroads," as it was named, was meant to test the effect of a nuclear bomb on a naval armada. Two tests were executed. "Shot Able" was an aboveground atmospheric explosion, while "Shot Baker" was an underwater detonation. Decommissioned and captured ships from World War I and World War II were placed in the ocean. Animals were strapped to some of the ships to test the effects of the radiation blast on living tissue.

The public was concerned about this testing, especially the underwater detonation. Admiral William H. P. Blandy, who was in charge of the operation, even gave a press conference during which he denied such possible outcomes as a hole being blown in the ocean floor, causing all the water to drain out the bottom.

The water surge that resulted from the Baker detonation didn't blow out the ocean floor, but it did drench the entire test fleet with contaminated water. The officer in charge of safety collected a large fish and placed it next to a sheet of x-ray film. The scales of the fish were so radioactive that it was able to produce its own x-ray. A subsequent test was cancelled, and the remaining ships were sunk and abandoned.

# Castle Bravo

In 1954, the United States conducted a nuclear test that would become its last. Castle Bravo was the codename given to this test of a new design—a thermonuclear hydrogen bomb. The expected yield of this design was to be 4,000 to 6,000 kilotons. This was more than 200 times greater than the bombs dropped during World War II.

The fuel for this bomb was a mixture of two isotopes: Lithium-6 and Lithium-7. Scientists had assumed that only the smaller isotope would be reactive. When the test was conducted, the actual yield was 15,000 kilotons, 2.5 times more than expected. The heavier isotope had also reacted.

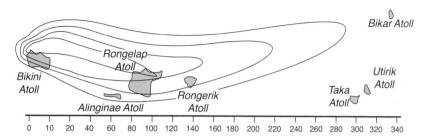

*Prevailing winds spread a plume of radioactive material for miles east of the Bikini Atoll, covering inhabited islands and an undetected Japanese fishing vessel, causing a great deal of radiation sickness.*

 **A LOOK BACK**

The Castle Bravo detonation and the subsequent contamination of the Japanese fishing vessel named *Lucky Dragon 5* caused a huge stir in Japan. Understandably, the country was still traumatized from the destruction of Hiroshima and Nagasaki. Popular culture responded to this anxiety with the movie *Godzilla*. The plot of the original film, released the same year as Castle Bravo, tells of a giant mutant sea monster attacking Japan. In the movie, the creature was discovered to be mutated by nuclear testing conducted by the United States.

The international concern about the amount of radioactive material spread from this test led to the eventual passage of the Partial Nuclear Test Ban Treaty. This was an agreement signed by the United States, Britain, and Russia that the countries would no longer perform atmospheric, surface, or underwater nuclear testing.

# Nuclear Energy

In 1953, President Dwight Eisenhower gave a famous speech to the United Nations called "Atoms for Peace." Consider these two lines from the speech:

> *"My country wants to be constructive, not destructive."*

> *"… the United States pledges before you … its determination to help solve the fearful atomic dilemma—to devote its entire heart and mind to find the way by which the miraculous inventiveness of man shall not be dedicated to his death, but consecrated to his life."*

President Eisenhower was marking a shift in the focus of nuclear research—away from designing more powerful weapons and toward finding a way to harness that energy for the benefit of everyone. In other words, generating electricity from nuclear fission.

## Pressurized Water Reactors

There are a few different kinds of nuclear reactors, each with its own design variations, and all incredibly complex. To keep things simple, we're only going to cover the most common design, called Pressurized Water Reactors (PWR).

The overall idea is actually a fairly simple one. Water is pumped through the area where the fission reaction is taking place. The heat from that reaction will cause the water to boil. As it boils it turns to steam, which passes through a turbine that spins a generator, creating electricity. If this all sounds familiar, it should—fossil fuel plants work basically the same way!

There are a few additional complications. Performing a nuclear fission reaction is much more dangerous than simply burning a fossil fuel. Fission reactions can quickly get out of control and release the energy too quickly. To prevent this, the fuel is continually bathed in water. This water is highly reactive, so it's kept separate from the water that spins the turbine (and is released into the environment as steam).

## Fuel

The fuel for nuclear power plants is Uranium-235, the same that was used in the first generation of nuclear weapons. The uranium is enriched to a level of 3 percent (a much lower level than weapons-grade) and made into pellets.

The pellets (1.5cm) are stacked end-to-end in long tubes called *fuel rods* (4.0m). About a hundred fuel rods are bundled together to make a fuel assembly. Thousands of fuel assemblies make up the *reactor core*.

One of the advantages of using uranium as a source of fuel is that it's very compact. One fuel pellet contains the equivalent amount of energy as 120 gallons of oil, 1 ton of coal, or 17,000 cubic feet of natural gas. In other words, you can generate a lot more electricity from the same amount of uranium than you could from any of the fossil fuels.

## Control and Containment

How do you control a reaction that has enough explosive energy to level an entire city? We already know that the nuclear core is continually bathed in water to help keep it cool. However, this alone is not enough.

Remember, the chain reaction works by producing neutrons, which collide with uranium nuclei, causing them to split and produce even more neutrons. Nuclear plants control the production of these neutrons by inserting material that actually absorbs them into the reactor core. These are called *control rods*.

 **DEFINITION**

Within a nuclear reactor, the uranium fuel is pelletized and stacked in long **fuel rods. Control rods** made of neutron-absorbing material are placed in between the fuel rods. These two materials together make up the **reactor core.** The core is surrounded by a **moderator solution** like water that keeps the temperature from getting too high.

When the control rods for a reactor are fully inserted into the core, enough stray neutrons will be absorbed to eventually halt the reaction. By raising or lowering the control rods, the output of the reactor can be altered based on the current demand for electricity.

The last and most important safety measure is the containment structure, the building that houses the nuclear core. Layers of concrete and steel are used to build this structure, which is several feet thick. The steel and concrete effectively absorb any stray radiation produced by the fission reaction. An individual could stand directly outside a containment building and receive no elevated amounts of radiation at all.

 **CASE STUDY**

One of the most common misconceptions about nuclear reactors is the fear that radioactive pollution is being released into the environment, either through air or water. A facility that's operating properly should release no radiation at all. What about the billowing smoke that arises from the huge cooling towers? This is simply excess steam released from the secondary loop of water that's boiled to spin the turbine. The water in this loop never comes into direct contact with the nuclear core, and is free of contamination.

# Nuclear Disasters

When the first nuclear reactors went online in the 1960s, optimism over the technology was very high. In fact, the chairman of the Atomic Energy Commission predicted that the electricity generated would be "too cheap to meter"!

Between 1970 and 1974, American utility companies ordered the construction of 140 new reactors. Electricity generated from these plants cost about half as much as that from burning coal. However, this optimism soon faded in the face of a few incredibly devastating accidents.

Of those 140 reactors that had been ordered in the early 1970s, 100 of them were subsequently cancelled. By the 1990s, electricity from nuclear power was actually more expensive than electricity produced by coal.

## Kyshtym Disaster

The arms race, the early rush to develop the most advanced nuclear technology as quickly as possible, was a dangerous one. This was particularly true in the Soviet Union, as safety and environmental concerns largely were not taken seriously. A big price was paid for this haste in a series of radiation contamination incidents at a nuclear fuel processing plant in Russia.

A series of six nuclear reactors was constructed at Mayak, each of which used a design that directly discharged cooling water back into the environment. Modern reactors, as you might recall, keep this water completely separate from water from the turbines. Waste was dumped into the nearby Techa River and Lake Karachay.

Worsening this contamination was an explosion that occurred in a radioactive waste storage tank in 1957. The radioactive plume that was released contaminated at least 22 villages, with even more radioactive material settling into the Techa and Lake Karachay.

The story gets still worse, particularly in regard to Lake Karachay. The region experienced a severe drought that dried up all the water in the lake. This exposed the radioactive material at the bottom to the wind, allowing it to become airborne. Eventually, the lake was filled with thousands of concrete blocks to keep the sediment permanently trapped at the bottom.

## Three Mile Island

The first major nuclear accident in the United States occurred in 1979 in the center of Pennsylvania. The Three Mile Island nuclear facility, built in the middle of the Susquehanna River, had four pressurized water reactors.

The accident started when a relief valve became stuck in the open position, allowing large amounts of coolant water to escape from the nuclear core. The operators of the plant didn't realize that the coolant level was dropping. As the coolant level dropped, the heat and pressure inside the core began to build up. One of the operators believed the pressure was due to an excess of coolant fluid, and it continued to drain out.

The coolant fluid, which is highly radioactive, leaked out of the primary containment structure, contaminating the rest of the plant. Meanwhile, the portion of the core that was not covered in coolant water melted. This is why Three Mile Island is referred to as a partial meltdown. Eventually coolant water was pumped back into the core, but not before a voluntary evacuation of the surrounding area had been conducted.

This near disaster terrified the public, especially because a movie with an eerily similar plot titled *The China Syndrome* had been released only a week before the accident. Federal regulation of nuclear power plants increased. The price of electricity from nuclear power increased, and the profitability of the plants fell. Far fewer were constructed during the following decades.

## Chernobyl

The first full meltdown of a commercial nuclear power plant occurred at the Chernobyl plant in what is now the Ukraine in 1986. This disaster was the result of a series of design flaws and operator errors.

The accident started when the operators were running an experiment to see how the backup generators would respond if an emergency shutdown had to be ordered. Backup generators are essential in a nuclear plant to keep the water pumps running to cool down the core in case normal electricity generation is interrupted.

A series of accidents and design flaws caused a power spike during the experiment. The core overheated and exploded. A buildup of steam pressure caused the entire roof of the containment building to be ripped off.

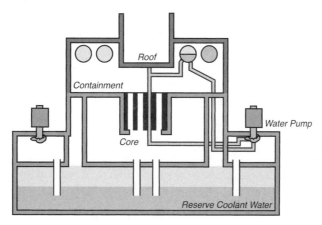

*The Chernobyl reactor had a few design problems that contributed to the meltdown disaster. The containment structure was insufficient, and a pool of reserve water was placed right below the core.*

A brigade of firefighters was quickly brought in to extinguish the fires. They were not equipped for the nature of the accident and were exposed to fatal levels of radiation. The burning nuclear core was eventually extinguished by dropping bags of sand, lead, clay, and boron into the reactor from helicopters.

**A LOOK BACK**

One of the firefighters who responded to the Chernobyl meltdown was asked if he and the others were at all aware of the danger they were walking into. "Of course we knew! If we'd followed regulations, we would never have gone near the reactor. But it was a moral obligation—our duty. We were like kamikaze." The firefighters, the operators of the plant, the pilots of the helicopters that dropped bags of sand into the reactor, and the cleanup crews that built the containment structure were all exposed to massive amounts of radiation. Most of them knew of the danger, but bravely continued in an effort to prevent the radiation from reaching the rest of the population.

The nearby employee town of Pripyat was not evacuated for more than 24 hours. When the evacuation was finally announced, it was labeled "temporary" and the residents left nearly all their belongings behind. Everything remains at that ghost town today, including carnival rides for a festival that would never be held.

The radioactive debris from the explosion was shoveled into what was left of the reactor and a huge concrete structure called "the sarcophagus" was built over the top of it to seal the radiation within.

The ecological impacts of the Chernobyl meltdown are massive. The nearby Pripyat River, which feeds into a reservoir that supplied 2.4 million people, was contaminated. Four square kilometers of forest turned reddish-brown and died. An exclusion zone of 19 miles was drawn around the reactor. Living in this zone or eating any food from this zone is considered unsafe. Rates of thyroid cancer, miscarriage, and birth defects have skyrocketed in the nearby populated areas.

## Fukushima

Fukushima Daiichi is a nuclear plant that was constructed in northeastern Japan. Like other power plants, it was built near a reliable source of water—the Pacific Ocean. The plant was constructed on a bluff about 35 meters above sea level, which had to be lowered to 10 meters so it could be anchored into a layer of bedrock. The plant was built with a series of sea walls, and was believed to be able to withstand even the worst earthquake or tsunami possible.

Unfortunately, the safety precautions were not sufficient. On March 11, 2011, a 9.0 magnitude earthquake centered nearby in the ocean triggered a massive 14-meter-high tsunami that overcame the bluff and sea wall designed to protect the plant.

While the plant was shut down in advance of the tsunami, fission reaction produces a great deal of heat that takes a long time to cool down. Water must be continually pumped through the plant while the temperature is lowering. A series of diesel generators were constructed to keep these pumps running, even in case of a power failure. These pumps were housed in the basement of one of the turbine buildings, and were also flooded by the tsunami. With the failure of the diesel generators, only a limited battery backup system remained. When the batteries ran out of power, the cooling system shut down, and a meltdown occurred in the reactors of the plant.

Like Chernobyl, a great deal of radiation was released. Some of it was airborne, from venting gas that had built up within the reactor. Some of it was in the water, as coolant water had to be ejected directly into the sea. The area immediately surrounding the plant is uninhabitable, and radioactive isotopes have been detected in food produced in the region.

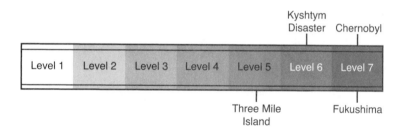

*The nuclear event scale rates each of the meltdown disasters based on the relative amount of radioactive material released and danger to living organisms.*

# Nuclear Waste

When a piece of wood or lump of coal is burned, energy, heat, and waste gasses are released. The noncombustible remnants, primarily minerals, are left behind as ash. The nuclear fission reaction is similar, except instead of an inert, nonhazardous end product, nuclear waste is hot, highly reactive, and very dangerous.

Uranium fuel rods reach a point where they no longer release enough energy and neutrons to sustain the chain reaction needed to power the reactor. At this point, the fuel rods are considered spent and, along with any other elements formed from the fission reaction, are high-level nuclear waste.

## Nuclear Decay

By their very nature, radioactive elements are unstable. The forces that normally maintain the integrity of a nucleus are insufficient for atoms of radioactive elements. Their nuclei are simply too large. As a result, stray particles of energy, protons, or neutrons are released. There are three common types of radioactive decay.

Alpha particles are the largest of the different forms of decay. They can be blocked by the dead outer layers of human skin and are harmless, as long as they are not ingested or inhaled. This is why workers in radioactive areas always wear dust masks! Since these particles are large enough to be blocked by skin and protective clothing, they're considered the least dangerous.

Beta particles are smaller and not always shielded by human skin. Again, these are most dangerous to human health when inhaled or ingested, allowing a direct route to the inner tissues of the body.

Gamma rays are not particles, but a form of energy with an extremely short wavelength. This enables them to penetrate skin and muscle easily, causing damage to inner organs. Gamma rays are even more penetrating than x-rays!

## Radioactive Half-Lives

While this tendency of radioactive elements to decay makes them dangerous, it also means they gradually break down into smaller (and more stable) elements. In fact, the rate of decay of any given radioactive isotope is measured as a half-life. This is the amount of time needed for half of a given amount of an isotope to decay. The fraction of material still radioactive will decrease exponentially, from $\frac{1}{2}$ to $\frac{1}{4}$ to $\frac{1}{8}$ to $\frac{1}{16}$, and so on.

So, problem solved, right? Take the spent nuclear fuel, allow it to gradually cool behind thick layers of concrete and steel until it's inert! Unfortunately, the rate of decay varies greatly for the different components of high-level waste. Iodine-131, for example, has a half-life of 8 days. Plutonium-239, however, has a half-life of 24,000 years.

Radioactive waste is generally considered "safe" to dispose of normally after 10 half-lives. Let's compare the relative length of time to do this for the two isotopes just mentioned.

Iodine-131, with a half-life of 8 days, will meet this standard after 80 days in storage. No problem.

Plutonium-239, with a half-life of 24,000 years, will meet this standard after 240,000 years in storage. Big problem.

To put this length of time in perspective, among the oldest standing man-made structures are the pyramids of Giza at about 10,000 years. Fossil evidence suggests that the *Homo sapiens* species has existed for about 200,000 years.

## Short-Term Disposal

Before we can tackle the problem of storing radioactive waste for thousands of millennia, we first need a short-term storage solution. Spent nuclear fuel doesn't generate anywhere near the amount of heat of a fully powered reactor core, but it still does release a significant amount. Fuel will generally be kept in water-cooled, sealed chambers for several years. This is known as wet storage. These pools cannot become too crowded, as the fuel still has the potential to reach critical mass.

This leads to the development of dry storage casks. These are large, cylindrical, aboveground structures that entomb the spent fuel rods in layers of concrete and steel. While these casks are adequate for shielding the gamma radiation from the outside environment, they certainly are not designed to last for thousands of years.

## Long-Term Disposal Methods

More than 40 years have passed since the first commercial nuclear reactors went online. In this time, no method or site has been agreed upon as a final resting place for the generated waste. A few ideas have been suggested, but each carries real danger and great risk.

One suggestion, for example, was to shoot the nuclear waste into space. It could either be directed toward the sun, where it would burn up harmlessly, or into the outer reaches of the solar system. This idea falters, however, when one considers the risk of a shuttle explosion. While there have only been two explosions in all the missions into space, the literal fallout if this were to happen with a payload of nuclear waste would be catastrophic. The exploded waste would spread throughout the atmosphere as dust, where it could be easily inhaled by living organisms all over the world.

The most agreed-upon consensus was to simply bury the waste. A simple landfill would not do, however. A location was needed deep underground, in a geologically stable (earthquake-free) area, with very little precipitation to corrode the metal instruments and storage containers.

An ideal location was found in the Yucca Mountain range in Nevada. This location is away from active fault lines and active volcanoes, experiences very little precipitation as a desert, and is 100 miles away from the nearest population center, Las Vegas. Tunnel construction began in 1994. Inexplicably, the residents of Nevada objected to these plans and fought them both in court and through the Nuclear Regulatory Commission. In 2009, plans to store nuclear waste in Yucca Mountain were cancelled.

Research is still underway to find a satisfactory final resting place for nuclear waste. Scientists believe there may be deep underground geologic formations that the waste can be permanently deposited in that would be geologically stable for millions of years. However, for the time being, nuclear waste remains a problem with no real solution.

## The Least You Need to Know

- Nuclear technology was developed first as a weapon, and was deployed twice during World War II.
- During the fission chain reaction, a nucleus of Uraniuim-235 is split, releasing heat and stray neutrons. When enough uranium is present, the reaction will be self-sustaining.
- The United States conducted extensive atmospheric and oceanic nuclear testing up until the Castle Bravo contamination incident, which resulted in the Partial Test Ban Treaty.
- Nuclear reactors work by using the heat of fission to boil water, which turns to steam and spins a turbine. Only steam is directly released.
- A nuclear meltdown occurs when the core is uncovered and overheats. The temperature of the fuel gets so high that the molten mass can actually melt through the containment structure.
- Nuclear waste is radioactive elements left over from the fission chain reaction. These must be contained until they naturally decay, which could take thousands of years.

# CHAPTER 16

# Renewable Energy

The last two chapters have focused on efforts to produce something we generally take for granted: electricity. This seemingly mundane household utility is the root cause of all the consequences we covered from fossil fuels and nuclear energy. Coal mine collapses, oil spills, natural gas leaks, and nuclear meltdowns—each of these is the result of our insatiable demand for electricity.

Not only do these sources have a high potential for causing environmental damage and health effects, but they're non-renewable. There's a finite amount of coal, oil, natural gas, and uranium ore in the Earth's crust, and there will come a point when it's too difficult or expensive to extract what remains. Electricity from these sources will be impractical or simply too costly.

None of these energy sources can be considered sustainable. They're abundant in the short term, but different sources must be explored if we're to continue our dependency on electricity. We need renewable sources of energy—ones that won't run out; ones that don't come with the same level of ecological damage.

## In This Chapter

* Harnessing solar energy for heat and electricity
* How living matter can be converted to fuel without waiting millions of years
* The benefits and debates surrounding wind and hydro power
* Using heat from deep within the Earth to generate electricity

# Solar Energy

One of the most obvious renewable sources of energy is one that living organisms have relied upon for billions of years: the sun. A massive amount of energy from the sun constantly hits the Earth.

At the top of the Earth's atmosphere, about 1,330 watts of energy from the sun strikes each square meter. With a little physics and geometry, we can figure out just how much energy this is across the entire planet: 1,330 watts per square meter $\times 4\pi r^2$ (the surface area of a sphere) equals $9.3 \times 10^{25}$ watts. That's over a septillion watts!

Why don't we simply tap into this vast amount of energy? The problem is, those septillion watts are spread out across the entire planet. Solar energy is very diffuse and difficult to collect.

There are a few different methods of collecting and converting solar energy into more useful, commercial forms. Some produce heat, while others produce electricity. The common link between all these forms of energy is that they're renewable and, once installed, pollution free.

## Passive Solar Heat

The easiest and most direct way to harness solar energy is to use it for heat. A significant portion of the energy requirement of our society is either forced air heat (such as from a furnace) or heated water for cleaning.

*Passive solar heat* gets its name from the lack of any moving parts in the system. Have you ever walked through a windowed room on a sunny day? It warms up quickly, much more so than other rooms with fewer or no windows. Glass allows heat from the sun in, but keeps it from escaping as quickly. Greenhouses work the exact same way.

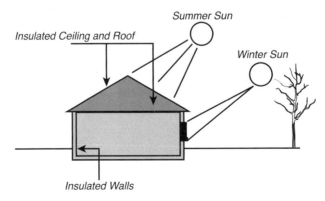

This seems like a trifling amount of energy, right? How could this possibly be used commercially? The challenge is to maximize the heat gain in the winter and minimize it in the summer. One way to do this is to construct the house in such a way that the maximum amount of sunlight will hit the windows during the winter, when the sun is lower in the horizon. Another way is to plant deciduous trees around the house. In the winter, the leaves will be gone and sun will shine through. In the summer, the tree will be in full foliage and provide shade to keep the house cool.

## Active Solar Heat

Passive solar heat can be helpful in reducing the amount of natural gas or electricity needed to produce heat, but it does nothing to deal with the demand for hot water. To harness solar energy in this way, some moving parts (water pumps) are needed. This type of system is called *active solar heat.*

 **DEFINITION**

**Passive solar heat** refers to a solar heating system with no moving parts. **Active solar heat** is a heating system that utilizes a pump to move water to a location where it can be heated by solar energy.

The focus here is water. First, the unheated water must be delivered somewhere it can absorb as much heat as possible from the sun as quickly as possible. An ideal spot would be the roof. Many systems use a series of small black pipes on the roof to collect the heat (black absorbs the most energy), which is then transferred to the water. The water, now warmed up, is pumped back to the hot water tank where it can be stored until needed.

There are a few important limitations to using passive and active solar heat. First, the sun is not 100 percent reliable—its energy is not available at night or during an overcast day. Second, each only provides heat. While this is an important part of our energy consumption, it's dwarfed by our need for electricity.

## Parabolic Solar Collection

Getting electricity from sunlight is a much more difficult affair. The energy transfer here isn't a direct one (heat-to-heat); it needs to be transformed. As you learned before, whenever energy is changed from one form to another, some will inevitably be lost due to inefficiency.

One method of making this conversion follows the same basic design of fossil fuel plants and nuclear plants: water is boiled. Now, a normal amount of sunlight certainly isn't enough to boil water. However, if enough sunlight is focused on one spot, this can work. The same idea is behind every little boy's favorite trick: burning paper (or insects) with sunlight and a magnifying glass.

*Parabolic solar collection* uses rows of mirrors arranged in such a way that sunlight from a large area is directed toward a central point. Some kind of substance, oil for example, is placed at that point to absorb the heat. This heat is then transferred to a water line, the water boils, the steam spins the turbine, and you have electricity. For this to be practical, you need a place with a lot of consistent sunlight, like a desert.

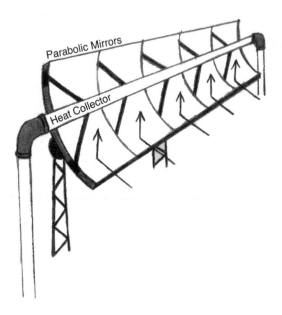

## Photovoltaic Cells

Finally, there are *photovoltaic cells,* more commonly known as "solar cells." Nearly everyone has experience with these on some level, as small cells are placed in some calculators to keep them running without batteries.

Solar cells work in four steps:

1. Particles of sunlight hit the solar cell and are absorbed.

2. The energy from the sunlight causes some electrons to be separated from the material in the solar cell.

3. The electrons begin to flow through the material, creating an electrical current.

4. The current is directed to the appliance needing the electricity.

Solar cells have been around for a long time, going all the way back to the nineteenth century. The problem is, they aren't very efficient and are expensive to produce. The efficiency is improving, with experimental cells reaching as high as 30 to 40 percent. However, the cost is still pretty high. Manufacturing practices in China have driven the cost down to $1.25 per watt, which is a huge improvement over the past but is still expensive when compared to the average cost of coal.

**DEFINITION**

**Parabolic solar collection** uses mirrors to focus sunlight to generate heat, steam, and eventually electricity. **Photovoltaic cells** are made of materials that absorb sunlight and directly generate an electric current.

# Biofuel

Most of the fuels we use, whether for heat or electricity, are burned. An alternative solution, then, would be to find something that is combustible but from a renewable source—one that comes from recently living organisms. In other words, biofuels.

The original biofuel is timber. Humans have burned wood for thousands of years as a source of heat. This arrangement worked fine when our population was small, but began to experience issues when the first large cities appeared.

A good example is Rome. The need for timber was so great in the city that all the forests around it were being clear-cut. Year after year, a larger and larger swath of land would be deforested. As the zone of deforestation grew, so did the distance traveled to find more forest. The trees simply could not regrow fast enough to keep pace with the demand of the city. This practice was not sustainable, and the Romans began experimenting with other sources of energy, including coal.

**CASE STUDY**

One of the major causes of environmental degradation in Somalia is deforestation and charcoal production. Regulations and law enforcement in the country are minimal at best, and a huge market exists in the nearby desert countries of the Middle East for the charcoal, as they do not have trees. If this problem continues over the long run, Somalia will lose its grazing lands to erosion and desertification, causing even greater levels of poverty.

## Dung/Biogenic Methane

Geographically, not all areas have access to timber. Any country in an arid region or a desert will not have any of that resource. Their heating needs will be much lower, but they still need to cook. What is their alternative? In some cases, the answer is dung.

The Native Americans who lived on the Great Plains had this problem. They turned to something called a "buffalo chip," a flat, dried piece of dung from a bison. Wood and coal were almost completely absent in the temperate grasslands of the area. This is another energy source that's limited by how fast it can be produced. A herd of bison is only going to produce a certain amount of dung, certainly not enough to support the energy needs of an entire society.

There is another way. Waste manure can be converted to a more conventional form of energy, methane, with the help of some bacteria. This is called biogenic methane because it's immediately produced by living things and not from geologic processes like underground methane.

Biogenic methane has a lot of potential, especially when you consider the amount of manure we produce on all our cattle feedlots and hog farms. The way production works is that the manure is added to a huge underground tank, along with some bacteria. Under the right conditions (sufficient moisture, warmth, and mixing), the bacteria decompose the manure and produce methane as their own waste product. The methane is piped out and collected, and can be burned directly for heat or as part of a small power plant. The only limiting factor is the huge initial investment of constructing and maintaining the fermentation system.

## Ethanol and Biodiesel

Biogenic methane is actually only one example of a waste product produced by a microscopic creature we can use for energy. Humans have been harnessing the power of yeast (microscopic fungi) to ferment plant products and produce alcohol for thousands of years. Up to this point, we have been mainly producing the alcohol for our own consumption and enjoyment. However, this alcohol is also very flammable, and anything that can be burned is a potential source of energy.

Have you ever visited a gas station that has labels on its pumps indicating that a certain percentage of the gasoline is ethanol? Ethanol is alcohol—no different than the kind we drink, except that it's 100 percent pure alcohol. Ethanol can be mixed with gasoline and used in any kind of internal combustion engine. It produces less pollution than gasoline, and avoids all the troubles associated with extracting and transporting crude oil.

Yeast can't just produce ethanol from nothing. It needs a food source—some kind of plant matter rich in complex carbohydrates. What do we currently use? What plant-based food do we have in abundance, especially in the United States? Corn! This seems like a possible solution to the problem of energy independence. We grow corn, ferment it into ethanol, and reduce the amount of crude oil we need to import.

The big catch with ethanol is that corn is a major source of food. By removing corn from the food supply system and using it as a fuel, food prices are likely to increase. These effects can be even greater in a drought-stricken year, as occurred in the Midwest United States in 2012.

"Political pandering comes in all shapes and sizes, but every four years the presidential primary brings us in contact with its purest form—praising ethanol subsidies amid the corn fields of Iowa."

John Sununu wrote this as part of an opinion piece in the *Boston Globe* in 2011. Iowa is considered an essential state en route to winning a presidential primary. As a result, few have the political will to consider reducing or eliminating government subsidies of ethanol production, even though most of the public would consider the costs of corn-based ethanol to outweigh the benefits.

# Wind

Power plants generate electricity through the turning of a turbine, which is essentially a really huge fan. Most power plants accomplish this through the movement of steam as it rises, but it's certainly possible to generate electricity through motion that doesn't involve heat. One way is with wind.

Humans have been harnessing the power of the wind for ages. Windmills, made of large wooden blades, would be connected to a shaft that would turn and grind up wheat to make into flour, pull up water from the ground, or perform some other task.

Wind power has a lot of potential benefits. Wind is a naturally occurring phenomenon, so it's certainly renewable. Nothing is burned, so no air pollution is produced.

## Wind Turbine Design

The most common design used for commercial wind turbines is called horizontal axis. These machines are fairly simple. A large set of metal blades is automatically pointed in the direction of the wind. As the wind blows through, the blades begin to rotate, which also spins a metal shaft they're all connected to. The shaft is connected to a generator, which converts this mechanical energy to electrical energy.

A single wind turbine will generate 1 to 7 megawatts of energy per year, depending on the design and placement. This alone is not enough energy to support a community of people. A large group of them must be constructed together. This is called a wind farm.

How big do these farms get? The largest is called the Gansu wind farm, located in northern China. In 2011, the facility produced about 6 billion kilowatt-hours of electricity. How much of a dent does this make in human energy demand? San Francisco, which has a population of about 800,000, uses about 18 million kilowatt-hours of energy per day. So this huge wind farm would supply the city for almost an entire year. The other good news is that the cost of producing this electricity is actually very comparable to that of coal. Wind power is cost effective!

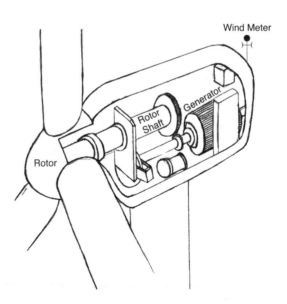

## Geography of Wind Power

The most obvious consideration when placing a wind farm is to find a location with a lot of wind. London, with its calm, foggy air, would be an incredibly poor choice. The Great Plains of North America, however, would work pretty well.

Scientists are using meteorological data to create models of what areas of the world would be the most ideal for wind farms. Some of the best spots are actually offshore, in the ocean. As you move farther away from land, you lose obstacles like mountains and trees that block wind. Other considerations include finding locations that are not exposed to intense storms (hurricanes, for example) and staying relatively close to populated areas.

**CASE STUDY**

Cape Wind is a proposed wind farm to be constructed in the shallow ocean near Cape Cod and Nantucket, Massachusetts. A great deal of controversy surrounded this project, involving concerns that ranged from effects on bird migration to destroying the natural aesthetics of ocean view properties. A coal and oil magnate even funded an opposition group against the project. Public support of the project prevailed, however, and it's approved to begin construction in 2013.

# Hydrokinetic

In the same manner that the energy of moving air can be converted into electricity, so can moving water. Hydrokinetic power, also called hydroelectricity, is one of the oldest and most prevalent forms of renewable energy.

Going back to 1920, about 40 percent of the United States' total electricity was generated by moving water. This number has since shrunk, mostly because the growth in hydroelectricity hasn't kept pace with the rapid increase in America's power consumption.

## Hydroelectric Dam Construction

Hydroelectric dams have become larger and larger over the last several decades. Here are some of the biggest ones in operation, with their power output in megawatts (MW).

### The World's Biggest Hydroelectric Dams

|  | Country | Year Built | Power Output |
|---|---|---|---|
| Hoover Dam | USA | 1936 | 2,080 MW |
| Grand Cooley Dam | USA | 1942 | 6,809 MW |
| Itaipu Dam | Brazil | 1984 | 14,000 MW |
| Three Gorges Dam | China | 2008 | 22,500 MW |

Conventional hydroelectric dams work in a very similar manner to wind turbines. As river water flows through the dam, it turns a turbine, which is connected to a generator.

Another type of dam, called pumped-storage, provides an opportunity to generate different amounts of electricity depending on the current demand. The way it works is by having a large water storage reservoir near the dam. Water is pumped from the main reservoir to this side one during the night, when electricity demand is low. When demand is high (during the day), the water is allowed to flow back from the storage reservoir, spinning a turbine on its way back.

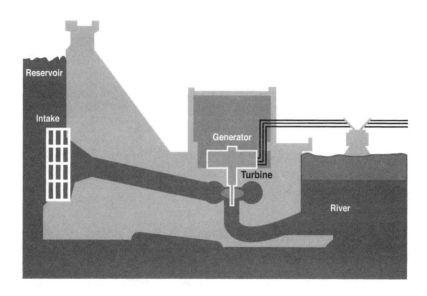

Hydroelectricity isn't limited to rivers. Tidal power stations are a form of hydrokinetic plant that uses the movement of water back and forth on the coast during high and low tide. While this movement is more predictable than solar and wind power, harnessing this energy so far has proven impractical. Most of the coasts with a great high/low tide differential are far from populated areas, and building and maintaining the plants is an expensive task.

## Ecosystem Effects

Hydrokinetic power seems to be a pretty flawless source of energy on the surface. No pollution is produced, and the reservoirs produced double as a reliable source of fresh water for consumption and irrigation. However, when a dam is constructed, the ecosystem of the surrounding area is changed in a dramatic way.

The most obvious effect is the submersion of the ecosystem under water. Dams are most often built where a river passes through a deep valley. Whatever biome was present in that valley is destroyed by flood. Depending on the size and location of the reservoir, people may be displaced.

**A LOOK BACK**

The Three Gorges Dam, now the largest hydroelectric dam in the entire world, was built among a great deal of controversy. Leading up to the dam's construction in 2008, the Chinese government had to relocate more than a million residents due to the flooding caused by the reservoir. A huge amount of forested area also was lost, threatening endangered species like the Chinese river dolphin and Siberian crane.

One hidden consequence of constructing dams involves the water itself. Dissolved in river water are tiny molecules of soil, nutrients, and partially decomposed matter. These are called sediments, and normally you can't see them unless a lot of soil erosion is occurring nearby (like after a heavy storm). The movement of the water keeps these sediments dissolved, or "suspended" in the water. When the water stops (in a reservoir), the sediments sink to the bottom. As a result, they build up behind the dam, leaving the water that flows out the other side much lower in nutrients. The water in the reservoir also warms, meaning it can evaporate much more quickly, causing the dissolved solids to concentrate even more.

Dams have other environmental effects. The migration of fish is effectively blocked, as they have no way of moving through the dam. Some dams employ a structure called a "fish ladder," which is a series of water-covered steps that the fish can travel up, allowing them to bypass the dam.

# Geothermal

Geothermal energy, like nuclear and coal, harnesses heat to produce electricity by boiling water. There is a big difference, however: geothermal energy uses naturally occurring heat from deep within the Earth. No fuel is burned.

A tremendous amount of energy lurks within the heat of the Earth. The total power generated by the movement of lava and magma is about double what our entire population uses annually! However, the problem with geothermal energy, just like solar, is that it's very spread out.

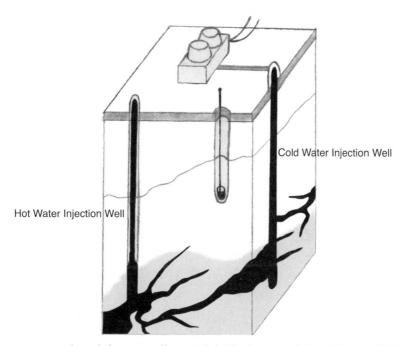

*To construct a geothermal plant, two wells are drilled. The first one, called the injection well, injects cold water down into the heated area. The second one, called the production well, directs the steam to the turbine. The steam condenses, and the cool water is again directed back into the Earth.*

The biggest restriction and drawback to geothermal energy is geology. Only the most geologically active "hot" areas have enough heat accessible to justify one of these plants. One good example of this is the Pacific "Ring of Fire"—all the coastal areas surrounding the Pacific Ocean.

## The Least You Need to Know

- Passive solar heating systems rely on orienting windows in such a way as to absorb heat from the sun during the winter.
- Active solar heating systems pump water to the rooftop, where it absorbs heat and is then used for bathing and cooking.
- Parabolic solar collection and photovoltaic cells both produce electricity from sunlight, but the efficiency is low and the costs are high.
- Biofuels, such as ethanol, biodiesel, and methane, are derived from living matter and considered renewable.
- Hydrokinetic and wind power work the same way: movement of water or wind spins a turbine, which generates electricity.
- Geothermal electric plants use steam from heated water taken from below the Earth's surface to spin a turbine.

# Consequences

In Part 5, we begin to look at some of the hidden prices we pay for our level of consumption, and the resources needed to support our population. Disease, toxins, and other human health issues have arisen as our population has grown and industrialized.

The following chapters specifically look at the degradation of our water and air. Water is one of the most basic necessities of life, but only a fraction of the Earth's water is available for us to drink and use to irrigate our crops. And air pollution has become an increasingly serious problem in a modernizing world. Next, we examine what is likely the greatest of all environmental consequences: global climate change. The warming of the Earth is a complex phenomenon. In these chapters I explain the basics behind the greenhouse effect and what causes it and also take a look back in Earth's history to get a sense of just how drastically the global climate has changed before. Then we focus on the human influences on the greenhouse effect, how the Earth is currently changing, and what this might mean for our society.

Finally, we see where it all ends. All the accumulated waste we produce, our garbage, has to be dealt with. The different options we have—from landfills to incineration—are compared, along with some more sustainable methods like recycling and composting.

# CHAPTER 17

# Human Health and Environmental Diseases

As we discussed in Chapter 9, the human population has grown exponentially since about the time of the Industrial Revolution. Life span has increased, infant mortality has dropped, and the overall quality of life in most of the world is vastly different.

Many of the treatments used before modern medicine were ineffective; reflecting a poor understanding of what actually made people sick. Bloodletting was one such practice. Doctors believed that sick people had an imbalance of their four "humors" or body fluids: blood, phlegm, black bile, and yellow bile. Draining some blood from the body was thought to help balance these humors and thus restore health.

There is a wide range of factors that can cause disease. Some are independently living organisms, others are chemical in nature, and still others are the result of lifestyle or genetics. In this chapter, we'll take a look at causes that can be directly traced to the environment in some way. In other words, diseases that are the result of something you are exposed to in the outside world.

## In This Chapter

- Microbes that can cause disease, and how they are treated
- Which infectious diseases are the most dangerous
- The different effects that a toxic chemical can have on human health
- How the risk level of different toxic agents is determined

# Biological Agents of Disease

One of the biggest catalysts of the increase in our life span was the advent of modern medicine. How much has changed in the last two centuries? Throughout the Middle Ages, most doctors and scholars followed the Miasma Theory of Disease. The basic idea behind this theory is that disease is caused by inhaling poisonous mist or vapor. Likely sources of infection were foul-smelling places; places where decomposition and decay were prevalent.

The Black Death was one of the most devastating epidemics ever to hit the human race. A third of Europe's population was wiped out. Following the Miasma Theory, doctors would dress in elaborate black robes and wear beaked masks to protect themselves from the infection. The beaks were filled with pleasant-smelling things, like dried herbs and flowers, to filter out the poisonous vapor.

As we now know, this theory was inaccurate. Diseases were mostly caused by microorganisms like bacteria or parasites or nonliving particles called viruses. It wasn't until the first microbiologists discovered this that the door was opened to immunizations, antibiotics, and sterilization techniques before surgery. While they are mostly invisible to us, there is a great deal of diversity in the microscopic creatures that plague humanity.

## Parasites

If you recall, parasitism is a symbiotic relationship where one organism benefits at the expense of the other. All living causes of disease technically fall within this category. The term parasites in this context actually refers to larger, multicellular organisms that can infect humans.

Human parasites come in many forms, covering many different classification groups. There are animal parasites, like the tapeworm. This organism spreads itself by depositing eggs in the host's intestines. The eggs are excreted through waste, then ingested (unintentionally) by another. Athlete's foot and ringworm are both fungal infections, spread by contact. Protozoa are a diverse group of mostly single-celled organisms that cause diseases like malaria and giardia.

 **CASE STUDY**

Malaria is a disease caused by a protozoan parasite—one that is spread through the bite of a tropical mosquito. This disease has plagued humans for so long that it has actually affected our evolution. Sickle cell anemia is a condition that results from a mutation—red blood cells take on an abnormal shape. While this carries a different set of health problems, it does grant anyone carrying the gene resistance to malaria. This is why the disease is so much more prevalent in those of African descent!

The one characteristic that each of these infectious agents has in common is they are *eukaryotes*. Each of these organisms is composed of cells that are fairly large, easily seen with a microscope.

While they are all dangerous, they are generally treatable with different medications. With the exception of malaria, these diseases don't really cause massive outbreaks.

## Bacteria

Bacteria are single-celled organisms that are much smaller than the cells of plants, animals, or fungi. Ironically, their small size actually gives them an added level of danger to multicellular organisms like us. They can slip past our body defenses more easily, multiply faster, and cause much more serious diseases.

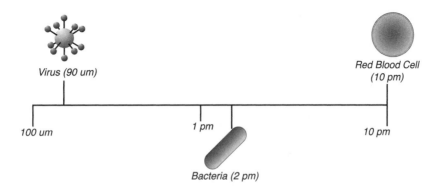

Virus (90 um)

Red Blood Cell (10 pm)

100 um

1 pm

10 pm

Bacteria (2 pm)

On the scale of microbiology, viruses are tiny infectious particles, much smaller than *prokaryotes* like bacteria, which themselves are much smaller than eukaryotic cells.

Every bacterial disease is different. Some are airborne, some spread through direct contact, and others require blood or sexual contact. Still others are passed into food or water sources through fecal contamination.

We tend to look at bacteria as malevolent creatures looking to kill us. That really isn't the case. Most bacteria are actually pretty harmless, and in the case of the ones living in your intestines right now, some actually benefit us!

 **DEFINITION**

A **eukaryote** is an organism made of cells that have specialized structures called **organelles.** Examples might include a nucleus, which contains the genetic material of the cell, or **mitochondria,** which generate the chemical energy needed by the cell. A **prokaryote** is an organism made of cells that lack a nucleus or any other membrane-bound organelles. Some prokaryotic cells are infectious, meaning they can invade organs and tissues made of eukaryotic cells.

Bacteria are simply looking for a warm, safe place to live with an ample supply of water and food. Unfortunately, their life cycle can wreak havoc on our normal body balance. Consider *Escherichia coli* O157:H7. If this bacterium takes up residence in your intestines, it begins producing a toxin. This toxin is a defense mechanism, meant to keep competing bacteria away. Unfortunately, the toxin has the side effect of interfering with connections between the cells of our intestinal wall. This allows water to leak into our intestines, causing diarrhea.

Bacteria, as dangerous as they can be, are generally treatable. Antibiotics are a group of chemical compounds that kill or interfere with bacterial growth. For example, one type of antibiotic blocks the bacteria's ability to form a protective cell wall around itself. Without this cell wall, the bacteria will burst and die. Even these medications fail, however, when the infectious agent gets even smaller and simpler.

## Viruses

Viruses are not considered living organisms. They're made of a small amount of genetic material surrounded by a protein coating. They don't have the ability to move, reproduce, or metabolize food on their own. They're completely reliant on other cells to perform these tasks for them.

The life cycle of a virus begins when it finds a host cell to infect. Most viruses are very specific in terms of the species they're able to infect and even the specific cell they target. The Human Immunodeficiency Virus (HIV), for example, only infects one type of white blood cell. The virus injects its genetic material into the cell, effectively taking over. The cell begins producing additional copies of the virus. Eventually, the new viruses spill out of the cell and spread, furthering the disease.

Viral infections are generally more difficult to treat than bacterial ones. Antibiotics don't work. There are antiviral medications, but they typically just slow down infections and don't actually kill the viruses directly. When it comes down to it, your immune system has to deal with the infection. This is why the common cold, the flu, and other viral diseases don't really have a cure.

## Prions

Viruses would seem to be the smallest, simplest, and most dangerous of the infectious agents. After all, they're essentially composed of two parts: the inner core of genetic material and an outer coating. Prions take this a step further. They're infectious proteins without any genetic material at all.

How could a protein by itself cause disease? Proteins have an unusual property: they change the shape of other proteins they come into contact with, rendering them unable to perform their normal function.

Prions cause a disease in cattle called bovine spongiform encephalopathy, or "mad cow disease." The prions infect and accumulate in the brain. Under a microscope, the brain tissue of an infected cow looks like it has holes (hence the term "spongiform").

Mad cow disease actually started in sheep. It migrated over to cattle when farmers in the United Kingdom began taking all the leftover parts of sheep (brain, spinal cord, bones), grinding them up, and feeding them to cattle as a protein supplement. Once the disease became prevalent in the cattle population, it made its way to humans. The only way for the disease to be spread is by ingesting infected nervous system tissue. Unfortunately, some of the brain tissue and spinal cords of the deceased cattle made their way into the meat people were buying.

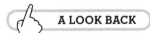 **A LOOK BACK**

The emergence of mad cow disease was a disaster for cattle ranchers in the United Kingdom. The disease can lie dormant in animals for years without displaying any symptoms. No definite diagnosis can be made without performing a brain autopsy. As a result, thousands of cattle were slaughtered and destroyed. The rest of the world banned all beef exports from the United Kingdom.

# Emergent Disease

Not all diseases are equally deadly or contagious. Some, like the viruses that cause the common cold, are a short-term annoyance. Others can be devastating. The Spanish Flu outbreak of 1918 killed 50 to 100 million people worldwide over just a matter of months!

Diseases like this, ones that our species has never previously encountered, are called emergent diseases. Humans have been on the Earth for tens of thousands of years. Our immune systems have grown and evolved to better combat disease. Emergent diseases present a special problem for us: our immune systems, for one reason or another, are not completely prepared to deal with these pathogens.

## Zoonosis

Emergent diseases often originate in other species. This might seem to contradict something we covered earlier—that viruses are very specific in what they infect. However, viruses also have the ability to mutate easily. They have very small genomes and reproduce incredibly quickly. Just by random chance, a mutation can occur that grants the virus the ability to be transmitted in a new way or infect a new organism or type of cell.

When one of these mutations allows the virus to move from a non-human animal species to us, it's called a zoonosis. Many of the emergent diseases that have plagued the human population in modern history are known to be zoonoses.

Remember the H1N1 "Swine" Flu epidemic in 2010? That was a virus that originated in a hog farm somewhere in Mexico. The disease rapidly spread through the population. Fortunately, it didn't cause anywhere near the death toll of the Spanish Flu. Nobody knows exactly where that flu started, but modern research suggests it likely originated from a swine or avian source.

**CASE STUDY**

A great deal of research has been conducted to determine exactly where the virus that causes AIDS came from. Most scientists now believe that the original host was chimpanzees in Africa. How did it cross over to humans? The most accepted theory is that it spread when natives hunted and killed chimpanzees for meat. Blood contact is inevitable when slaughtering and preparing an animal. The virus probably mutated, gained the ability to infect human cells, and entered our population.

## Antibiotic-Resistant Disease

Emergent diseases are dangerous due to their newness to our species. Other infectious agents have been able to evolve ways to elude modern medicine.

Antibiotic-resistant bacteria are just that. They're either immune or less susceptible to at least one antibiotic. Like viruses, bacteria are small and have very small genomes. A small mutation can make a big difference in the characteristics of that bacterium.

Bacteria also have the ability to trade bits of their DNA, called *plasmids*, with each other. Plasmids are separate from the main DNA strand of the bacteria. They can easily be replicated and passed along in a colony. Additionally, they reproduce asexually.

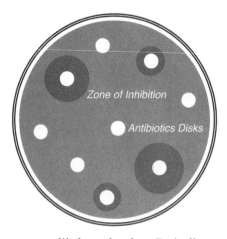

Where is antibiotic resistance most likely to develop? Basically, anywhere antibiotics are being used heavily. In other words, the more often antibiotics are prescribed, the more likely the bacteria are to develop a gene to resist them. Mutations are random, but given enough chances, resistance will emerge. What are some ways we use and overuse antibiotics that lead to this problem?

- Taking antibiotics for a nonbacterial infection.

- Providing antibiotics to farm animals.

- Not finishing a prescription of antibiotics.

The biggest breeding ground of antibiotic-resistant bacteria is hospitals. Think about the sheer number of sick people that pass through and the amount of antibiotics prescribed. When a person contracts a bacterial infection in a hospital after a surgery, it is almost certain to be resistant to at least one antibiotic, and often more.

# Environmental Toxins

Not all diseases and health conditions are the result of biological infectious agents. Sometimes, exposure to a chemical can have serious health consequences.

The effects that a specific toxin has on an organism vary widely. Allergens, for example, activate the immune systems of those who are sensitive to them. Examples include food allergens, such as peanuts, or airborne ones like pollen or pet dander. Not everyone is equally sensitive to these allergens: some contract mild symptoms like watery eyes, while others experience life-threatening swelling, or anaphylaxis.

This range of effects is similarly seen in many different environmental toxins. This is one of the aspects of this field of study that makes it so complex. If a person gets sick, how do you isolate the specific compound that caused it? Is this an individual sensitivity? Does the toxin interfere with normal body processes or cause an overreaction like allergens? Are the effects of the toxin acute (short-term) or chronic? Let's take a look at some of the major categories of toxins to get a sense of what is out there.

## Neurotoxins

Neurotoxins are a special class of poisons that specifically affect nerve cells. The single most prevalent type of neurotoxins is heavy metals. Examples include lead, mercury, and cadmium.

The actual symptoms experienced as a result of the neurotoxins vary. Lead, for example, is known to lower IQ, especially in young children.

**CASE STUDY**

Ancient Rome was one of the largest and most advanced cities of its time. The collapse of the Roman Empire is a heavily studied branch of history. One relevant theory actually centers around lead exposure. Romans used a compound called lead acetate as a sweetener, especially in wine. Continued exposure to lead is known to lower IQ. Only the nobility would have had access to that much wine, leading to the intriguing possibility that the upper class was experiencing lower than normal IQ levels!

Mercury, on the other hand, tends to have more psychological effects. Remember the Mad Hatter from the *Alice in Wonderland* book? The backstory behind that character is that his erratic behavior is the result of exposure to mercury. Indeed, at one time mercury was used to separate animal fur from skin to eventually make hats.

## Hormone Disruptors

Some toxic chemicals resemble normal compounds used by the body. One such example is hormone disruptors. These are toxins that either mimic or interfere with hormone functions in the body.

The single biggest source of exposure we have to hormone disruptors is plastics. Two examples are BPA (Bisphenol-A) and phthalates. BPA is a compound added to some plastics and plastic linings. Pthalates are added to plastics to make them softer and more flexible.

The hormones that are most often affected are estrogen and testosterone—the sex hormones. There are a lot of hypothesized links between exposure to these hormone disruptors and certain reproductive issues in the human population:

- Decreased sperm count and motility in men.

- Increased male birth defects.

- Early-onset puberty.

- Increased rates of testicular and breast cancer.

None of these links are definitely proven. Performing long-term studies on the effects of specific chemicals on the human population is very difficult and time consuming. As a result, some countries have outright banned these compounds, while others (such as the United States) continue their use.

# Mutagens

*Mutagens* include anything that can damage or alter genetic material—the DNA found within the nucleus of a cell. Your DNA is well protected in your body. Not only is it kept within a special compartment in your cells, but the outer layers of your skin are dead cells that can't be mutated.

In order for an agent to damage DNA, it must be able to somehow penetrate these defenses. Certain wavelengths of energy have an extremely high frequency, one that allows them to penetrate living tissue. A good example of this is X-rays, which pass through your soft tissue but not your bones. Ultraviolet radiation from the sun is a mutagen, as is nuclear radiation.

Other mutagens must enter the body before they can have these effects. Alcohol, for example, when ingested by a pregnant woman, can pass through the umbilical cord and have teratogenic effects on the fetus. Cigarette smoke is a carcinogen and can mutate the cells lining the mouth, throat, and lungs.

 **DEFINITION**

A **mutagen** is a chemical or type of energy that can damage the DNA of living cells. **Teratogens** are mutagenic agents that are known to cause birth defects. **Carcinogens** are mutagenic agents that cause cancer.

# Persistence and Bioaccumulation

When it comes to toxic compounds, one of the greatest areas of concern is when they contaminate water. All living things require water to live. Any toxins present in water can become incorporated into living tissue and build up over time. This is called *bioaccumulation.*

The story doesn't end there. Exposure to a contaminant by drinking or breathing water (if you have gills) is only the beginning. Organisms at the bottom of the food chain accumulate small amounts of the toxin. When they're eaten, this toxin is passed along. The problem is, organisms higher on the food chain have to eat a lot of smaller organisms to survive. Each meal increases their exposure. This continues to build all the way up the food chain, in a process called *biomagnification.*

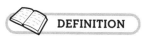 **DEFINITION**

**Bioaccumulation** is the gradual build-up of a toxic substance in the tissues of an organism due to continued exposure through food, water, or air. A specific example is **biomagnification,** an increasing concentration of a toxic substance through a food chain, peaking at the top levels. Toxins that exhibit persistence, the property of not being easily broken down in the environment once released, are ones most likely to accumulate.

The contaminants that are most likely to cause this are persistent. They don't break down easily in an environment once released. Bioaccumulation and biomagnification don't happen overnight; the contaminant may be present for years before its effects on the top consumers are felt.

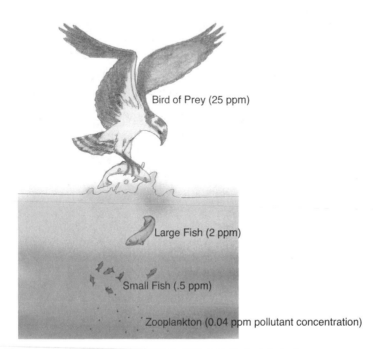

*Environmental toxins begin to accumulate at the bottom of a food chain, but grow to greater and greater concentrations as more predation occurs.*

## LD$_{50}$ and Toxicity

The world is starting to look like a dangerous place, right? We are all exposed to mutagenic agents, endocrine disruptors, and persistent neurotoxins. This exposure is inevitable; it's simply not practical to build yourself a sealed little bubble to live in. How, then, do we determine which of these toxins are the most dangerous? How does a government agency like the EPA decide how much of any one compound can be safely present in our water or air?

One such comparison tool is called LD$_{50}$. This is a measurement of the dose of any specific toxin that is required to cause death (lethal dose) in half of a given population.

This measurement is made by exposing a population of lab animals (generally mice or rats) to different doses of a toxin. The animals will all be the same species, have similar genetics, and eat identical diets to eliminate any outside variables. Data on dosage and death rate is collected and graphed.

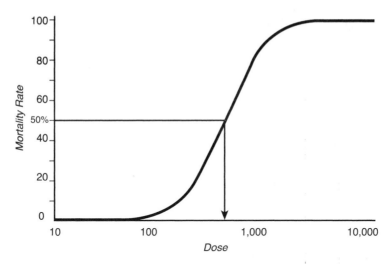

*The point at which the mortality rate for a given toxin is measured to be 50 percent is considered the $LD_{50}$ value.*

This data is useful in evaluating what the danger level is of any individual substance. For example, sugar has an absurdly high $LD_{50}$ level of 29,700mg/kg. Arsenic, on the other hand, has a much lower $LD_{50}$ level of 763mg/kg. The lower the measured $LD_{50}$, the more dangerous the substance is. One of the most dangerous of all is a waste product of nuclear fission: polonium-210. It has an $LD_{50}$ level of 0.000001mg/kg!

## Risk Management

We now have a lot of information: data about lethal doses and research into the specific effects of each toxin. The final question is, what should we really be concerned about? If you were an elected official in charge of improving the health of your constituents, what would your priorities be?

Answering this question requires a little risk assessment. Let's consider what risks are generally considered acceptable:

- Risks with a high probability of exposure with a low severity.

- Risks with a low probability of exposure with a high severity.

Canned tuna contains small amounts of mercury. This is widely known, yet tuna remains one of the most consumed fish species in the world. The amount of mercury is low, and it probably won't have any serious health effects (unless you really eat a lot of tuna). Nuclear power plants rarely experience meltdowns. Recall from Chapter 15 that there have been just four meltdown incidents in over 50 years.

So what are we likely to be exposed to that is known to have serious health effects? According to the U.S. Safety Council, the most likely cause of death for someone in the United States is heart disease. Heart disease is known to be linked to lifestyle choices, including high-calorie diets and little physical activity.

Heart disease and obesity are so commonplace, so mundane, that we often overlook them. Yet being overweight shortens the average life span by six years. Someone who truly understands risks like this would be much more concerned about their daily morning pastry than the pesticide residue that might be on a conventionally grown apple.

## The Least You Need to Know

- Infectious disease is caused by the spread of biologic agents like bacteria and viruses.
- Emergent diseases, the ones that spread the fastest, are often the result of mutated strains that originated in other animal species.
- Due to their small size and rapid growth, viruses and bacteria can quickly mutate and become resistant to medications we use.
- Each toxic substance has its own effects on the human body, ranging from mutating DNA to interfering with hormones to disrupting nerve function.
- The risk level associated with any toxic substance depends on the dose required to cause health effects, and the likelihood of an individual being exposed to it.

# Water Pollution

Of all the Earth's resources, water is the one that every known living organism needs to survive. When astrobiologists study distant planets to see if they can support life, they look for water.

Water, like other forms of matter, is continuously recycled throughout the Earth. The water you drink today has probably flowed down a river, evaporated from the ocean, and been consumed (and excreted) by countless other organisms.

Water is a renewable resource. We will never run out of it. However, there is a limited amount of fresh water we can drink. Should that water be lost or contaminated, our very survival is at stake.

## In This Chapter

- The major compartments of water on Earth
- What water sources people depend on the most
- What contaminants are found in fresh water
- The increasing contamination of the world's oceans

# Earth's Water

The Earth is mostly water. A quick look at any map or globe will make that clear. The oceans are incredibly vast, giving the illusion that water is an endless resource. Remember, however, that ocean water is salty and not readily available for our use.

About 97 percent of the Earth's total water budget is in the oceans. This leaves a scant 3 percent as fresh water. Where do you think the biggest compartments are after that? By far, the majority of the fresh water on the Earth is frozen as glaciers, icebergs, or sea ice. Most of the rest is underground.

Only the tiniest fraction, about 0.02 percent, makes up the water compartments that we think of most often: rivers, lakes, and the atmosphere.

## Water, Water, Everywhere

Water doesn't just stay locked in place. It travels around the Earth constantly. To understand this hydrologic cycle, let's follow a drop of water in the ocean.

As ocean water is warmed, the parts near the surface may *evaporate,* or move upward into the atmosphere as water vapor. This actually happens with all above-ground water sources. Even trees lose water this way, through a process called *transpiration.*

So our water drop has become vapor, and is now part of the lower atmosphere. The prevailing winds carry it across the ocean and eventually over land. At some point, the warm, moist air will *condense* (maybe as it travels up a mountain), and the now-liquid water will fall back to the ground as precipitation.

The next step depends on where our water drop lands. If it ends up on land, it will *infiltrate* down into the soil or flow as *runoff* into a river, lake, or ocean. At some point, evaporation or transpiration will return the water drop back to the atmosphere, and the cycle continues.

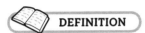 **DEFINITION**

Liquid water enters the atmosphere directly through the process of **evaporation.** This also occurs in plants, and is called **transpiration.** Eventually, the water **condenses,** or becomes liquid, and falls back to the Earth in the form of precipitation. The process of water soaking deep into the soil is called **infiltration.** When it moves over the soil and into a body of water, it's **runoff.**

## The Underground Reservoir

What happens to the water that infiltrates through the soil? To understand groundwater, we need to do a little digging. When you first start digging in the ground, you remove a shovelful of soil, place it aside, and a small, empty hole remains. This is called the *zone of aeration*, where the empty spaces in soil are mostly air.

What happens if you keep going? Eventually, you'll get to a point where the bottom of the hole keeps filling in with water, no matter how much you remove. You've reached the *zone of saturation*. These two zones are separated by an imaginary line called the *water table*. When someone digs a well, they need to at least go past the water table.

Some areas have an additional layer of water even farther down. *Aquifers* are layers of porous rock, sand, or gravel that are filled with water. They're isolated from the other parts of groundwater.

Wells deep enough to tap into an aquifer are called artesian wells. Aquifer water is considered some of the cleanest, highest-quality water available, because it's protected from activity on the surface by layers of rock.

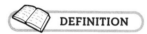 **DEFINITION**

The **zone of aeration** is the upper layer of soil, where the empty spaces within are most often filled by air. The **zone of saturation,** deeper down, has soil with empty spaces filled by water. The border between these two zones is called the **water table.** If an isolated source of water exists farther down below the zone of saturation, it's an **aquifer.**

# Depleting a Renewable Resource

In terms of meeting human needs, part of our population will be able to extract water from a nearby lake or river. Everyone else, though, will be reliant on groundwater.

To assume these sources of water are infinite is foolish. Yes, water is a renewable resource, but compare the rate of precipitation and infiltration to how much we withdraw. Once again, we have a problem with unsustainability.

What's the root of this overconsumption? The largest share of water use worldwide is actually agriculture. Irrigating crops is extremely water intensive, and nearly three quarters of what we withdraw is for this purpose. Remember, irrigation was one of the big changes of the Green Revolution. About another 20 percent of water is taken for industrial purposes, such as cooling water in power plants.

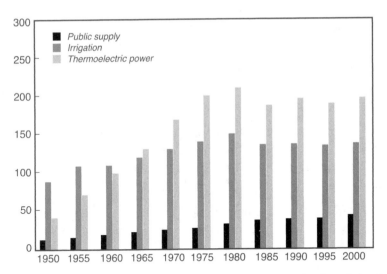

*Irrigation increased through the 1980s, reflecting changes brought by the Green Revolution. Public water needs also grew during that time.*

## Groundwater Overconsumption

Currently, there are more than 45 countries in the world experiencing something called water stress. This happens when the basic water needs of the citizens cannot be met. The countries experiencing the worst problems are, not surprisingly, arid ones such as in North Africa and the Middle East.

However, arid countries aren't alone in dealing with this problem. Any region that withdraws a lot of water, particularly for irrigation, is susceptible to shortages.

When water is withdrawn from the ground faster than it can be replaced, the water table actually drops. This is called a cone of depression.

This creates a lot of problems. First, any shallow wells that have been dug in the area will begin to run dry. Second, all the vacant, empty spaces created in the zone of saturation can actually collapse, forming sinkholes.

If the aquifer is located near a coast, the lowering of the water table can have the additional effect of drawing seawater in (since the sea level would be higher). This is called saltwater intrusion. Once saltwater starts to seep into an aquifer, it will eventually become too salty to be useful for irrigation.

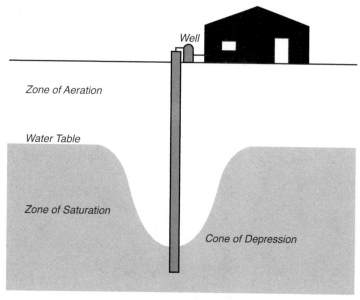

*Rapid withdrawal of groundwater from a small area can result in a rapid drop of the water table, creating a cone of depression.*

## Shrinking Lakes

Clearly, groundwater is a precious resource. What about our other major water sources? Lakes, as large and vast as they may appear, can still be subject to overconsumption.

In 1947, Joseph Stalin proposed a series of plans to help the Soviet Union avoid a repeat of a famine that had killed nearly a million people. One of those plans was to build a network of irrigation canals diverting water from two rivers into a desert region. The idea was to provide a dependable area of farmland that would not be at the mercy of Mother Nature.

The canals were not well constructed. They leaked from below and were uncovered, leading to evaporation. Much of the diverted water never made it to the irrigation fields. Meanwhile, the lake that these two rivers fed into began to shrink. The Aral Sea, as the lake was called, was once large enough to fill both Lakes Erie and Ontario. Today, only two much smaller portions of it remain; the rest is a dried lake bed.

 **A LOOK BACK**

The Aral Sea disaster did more than destroy a massive lake. The little water that remained was highly concentrated with sediments. The salinity was greater than ocean water. The exposed bottom of the lake contained pesticides, herbicides, and other industrial pollutants that had been previously dumped and forgotten.

Rivers aren't limitless, either. Many of the world's rivers are fed by glaciers. As we'll learn later on in Chapters 21 and 22, most of the world's glaciers are shrinking. If the glaciers were to completely disappear, the rivers they feed would also die.

## Water Conservation

At this point, you get the picture that more effort needs to be made to conserve water. Only about 5 percent of our water withdrawal is for household use, yet it's the one aspect that we as individuals have the most control over. With that in mind, what are our biggest domestic uses of water?

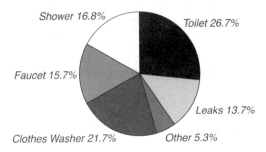

*The majority of household water use actually stems from just a few sources, primarily toilet flushing, clothes washing, and bathing.*

If we improve the efficiency of toilet flushing, bathing, and dishwashing, this would have a huge impact on our domestic water use. Each of these is a necessary part of living, so how can we change?

Many appliances are made to reduce water consumption. For example, a typical showerhead uses about 2.5 gallons of water per minute. A water-saver version uses 1.5 gallons per minute. If you take a 10-minute shower every day, that adds up to:

10 gallons per day × 365 days = 3,650 gallons per year!

Imagine how much would be saved if everyone made this switch! What about toilets? Newer models, called dual-flush toilets, have a half-flush button for liquid waste. There are also high-efficiency washing machines and dishwashers that use significantly less water.

# Polluting a Renewable Resource

Overconsumption of fresh water is bad enough; what if the water were to be degraded or contaminated? Unfortunately, it is. Water pollution occurred practically unchecked for nearly a century before the Clean Water Act and the Safe Drinking Water Act were written and signed.

Water pollutants are categorized based on where they originate. If a single location is discharging the pollution, then it's considered a point source. If there are multiple locations spread out across an entire region, it's a nonpoint source.

Examples of point sources would be factories, power plants, and oil wells. While each of these has a high potential for producing water pollution, they at least can be easily monitored and regulated.

Nonpoint sources are a lot more difficult. For example, if you had a lake that was becoming contaminated with herbicide runoff, where would you lay the blame? The nearby golf course? The soybean farmer upstream? The houses surrounding the lake with suspiciously weed-free lawns? Each of these sources is probably contributing to the problem, but managing and regulating all of them is a logistical nightmare.

## The Ones That Don't Go Away

The Clean Water Act was passed in 1972. Surely, in the decades since then the pollution that was dumped before has cleared up, right? In some cases, this is true. There are some pollutants, however, that simply do not go away.

Persistent toxins, as you might remember from Chapter 17, don't break down easily in the environment. They might be elements that simply cannot break down any further. Heavy metals like mercury and lead are persistent in water. They might also be complex synthetic chemicals that no decomposer has the ability to metabolize. Some organic industrial waste like polychlorinated biphenyls (PCBs) fit this description.

Either way, once these pollutants enter the water supply, they stick around. They bioaccumulate in the producers and primary consumers in the food chain—organisms like plankton and small fish. Eventually, these toxins biomagnify to dangerously high levels in tertiary consumers. Bottom-feeders are especially vulnerable, as many of these pollutants tend to concentrate there.

 **CASE STUDY**

The states surrounding Lake Michigan have developed a guide to what fish can be eaten from the lake safely and how often. The species of fish that are not recommended for consumption are large lake trout (top of the food chain), carp (bottom-feeder), and channel catfish (bottom-feeder). These fish are contaminated with PCBs—industrial chemicals that were legally dumped into Lake Michigan for years before the Clean Water Act was passed. These chemicals are known carcinogens and hormone disruptors.

## Too Many Nutrients

Lead, mercury, and PCBs; all these terrifying chemicals came from point sources and are no longer allowed to be dumped in water. What about the nonpoint sources?

People who contribute to nonpoint pollution aren't doing it intentionally. They might be applying some fertilizer or herbicide to their lawn or farm, and some of it runs off after a heavy rain. The amount of runoff from any one person's property is pretty small, but it adds up. Nutrient pollution primarily comes from these sources.

There are two main components of nutrient pollution: nitrates and phosphates. These are two of the three main components of fertilizer. What happens when they end up in water? It seems they stimulate growth of not only grass and soybeans but algae as well. If you've ever seen a lake or pond with a really thick layer of algae on top, you know what the end result looks like.

 **CASE STUDY**

Red tides are blooms—sudden explosions in a specific red-colored species of algae. These algae have serious effects on human health, as they produce a toxin that can infiltrate into shellfish we eat. To an extent, red tides occur naturally; but they've increased in size and frequency with the growth of nutrient pollution from coastal populations.

Algae are beneficial in small amounts. They perform photosynthesis and contribute to the overall oxygen content of the water. Too much can actually smother out larger organisms. Fish need to breathe oxygen from water, just like we do from air, and they can't breathe through a thick layer of algae.

Additionally, those algae will eventually die. As they decompose, the bacteria that feed on them will use up much of the oxygen in the water. The results are similar to what happens when raw sewage is introduced into a water source.

## Raw Sewage

Of all the different possible contaminants of water, raw human sewage seems pretty tame. Certainly, pouring raw sewage into our water supply might sound gross, but at least it is degradable.

That ability to be degraded, however, creates a whole new problem for aquatic ecosystems: dead zones, low-oxygen areas where no fish or other large organisms are found.

When raw sewage is dumped into a body of water, it carries a lot of coliform bacteria in it. These bacteria love sewage; it's a great food source for them. Just like lots of other organisms, the bacteria use the *dissolved oxygen* in the water. So you have a bacterial population explosion, followed by an *oxygen sag.* Large animals like fish need a lot more oxygen than bacteria do, so they have to leave or suffocate. The end result is a *dead zone.*

**DEFINITION**

**Dissolved oxygen** is the amount of oxygen gas present in a given sample of water. Biological oxygen demand is a measurement of how much of this oxygen is being consumed. An **oxygen sag** occurs when the demand is too great and the oxygen levels are too low. This creates a **dead zone,** a section of an aquatic ecosystem where no large organisms are found due to the low oxygen levels.

The good news about dead zones is that they will recover, in time. If a dead zone or oxygen sag occurs in a river, areas farther downstream of the discharge point will be just fine.

The dissolved oxygen levels immediately following the discharge of raw sewage drop rapidly. While they eventually recover, the life that can be supported in that portion of the water is dramatically altered.

To eliminate this problem altogether, we need to ensure that raw sewage is not directly discharged into water. It must be cleaned or treated first.

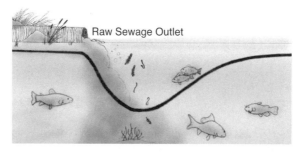

*Immediately following the dumping of raw sewage into a water source, the oxygen levels quickly drop. This limits the life that can survive in that water source.*

# Cleaning Up the Waste

The idea behind water treatment is a pretty straightforward one. When we flush the toilet, drain the sink, or run a load of laundry, a lot of water goes into the sewer system. This water contains a wide variety of contaminants, but here are the major ones:

- Organic waste
- Coliform (fecal) bacteria
- Phosphates
- Nitrates

The source of the organic waste and bacteria is pretty obvious. These contaminants must be removed not only to protect human health, but to prevent bacterial blooms (and drops in dissolved oxygen) from occurring. What about the nitrates and phosphates? A big source of phosphates is the soap we use. Laundry detergent, shampoos, and hand soaps often contain phosphates that help in dissolving and cleaning oils and grease. Nitrates are a common contaminant in runoff where any soil fertilizers are used.

## Primary Treatment

The first step in the water treatment process is to collect the wastewater in a big tank, where the components are allowed to settle. The heaviest components sink to the bottom and form sludge.

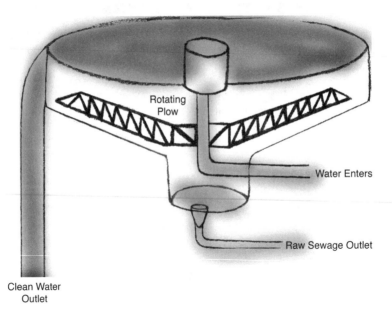

*Primary sewage treatment provides a means for the solid components to settle, resulting in an outflow of cleaner water.*

Sludge is mostly solid human waste and must be treated separately. Some communities use anaerobic digestion, essentially employing bacteria to break down the solid waste. This produces methane, which we defined in Chapter 16 as a biofuel. Other communities will compost the sludge into a usable fertilizer.

About half of the suspended solids (a polite term for human waste) are removed by this process. If the wastewater were to be returned to a body of water at this point, it would still have a sizeable (albeit reduced compared to no treatment) impact on the dissolved oxygen levels.

## Secondary Treatment

The secondary treatment process is also called biological treatment, because it takes advantage of sewage-loving bacteria to further reduce the suspended solids.

Some of the suspended solids are converted into gases, like carbon dioxide and hydrogen sulfide. Any remaining material is either settled out or filtered out.

When this process is used, a disinfection step is also necessary. Chlorine, ultraviolet light, or ozone gas is added to the water to kill off as many of the microorganisms as possible. The effluent (discharged water) will have a vastly reduced amount of organic waste, somewhere on the order of 90 percent or greater.

## Tertiary Treatment

Tertiary treatment is an additional step that mainly seeks to remove the nutrient wastes, the nitrates and phosphates. Nitrates are removed with the help of a species of bacteria commonly found in soil that performs "nitrogen fixing." Phosphates are also removed by bacteria, a species that accumulate the nutrients within their cells.

Once the bacteria have completed their work, the solids that have been produced can be removed from the effluent. The resulting discharge should have a minimal impact on the aquatic ecosystem that receives it. Dead zones, eutrophication, and the other side effects of dumping waste are largely averted.

# The State of the Ocean

In 2010, a gray whale beached itself and eventually died in West Seattle. Marine biologists studying the whale discovered an unsettling amount of garbage in its stomach. Contents included:

- Sweatpants

- A golf ball

- Surgical gloves

- Plastic bags and other plastic pieces

The majority of the contents found within the whale's stomach was algae, its normal diet. However, the garbage that had accumulated is telling of a greater problem: ocean pollution.

Most whales are filter-feeders. They take in massive amounts of ocean water and then push it out through baleen, which filters out any solids.

There are a lot of different contaminants found in the ocean, including many we've already covered, like mercury. Here we'll focus on two: oil and plastic.

## Oil Pollution

The release of crude oil into the ocean, given our appetite for the resource, is practically inevitable. We covered some of the worst spills in Chapter 14. The *Exxon Valdez* and the Deepwater Horizon oil spills would each be considered a point source of the pollutant. The effects on the nearby ecosystems were devastating, due to the sheer volume of oil released.

Yet oil spills like these are not the single biggest contributor to what is in the ocean. Nonpoint sources, leaking cars or people improperly dumping oil down their drains, are the biggest culprit.

Oil has different impacts on aquatic ecosystems, depending on the concentration present. Heavy amounts of oil are acutely deadly to most aquatic organisms. Smaller amounts are less so, although they can still biomagnify up a food chain.

The good news about oil pollution is that it will decompose over time. Crude oil is a naturally occurring substance, and there are species of bacteria that can break it down. Recovery may be slow, but it will happen. The same cannot be currently said for plastic pollution.

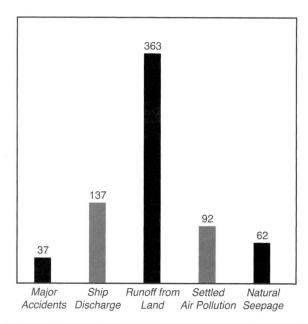

*According to the National Resource Council, the biggest source is natural seepage, followed closely by small discharges from ships or other facilities on land.*

## Plastic Pollution

Plastics are synthetic compounds made from products of crude oil. There are many different types of plastics, but they all have one thing in common: they're not found in nature. There aren't any organisms capable of degrading and breaking down plastics. So a plastic bottle in the ocean will slowly break apart into smaller pieces due to exposure to the elements, but it will never truly decompose.

As a result, there are areas in the ocean that contain more plastic than plankton. A major example is the Great Pacific Trash Vortex, located within a series of currents between the United States and Japan.

 **CASE STUDY**

Glass Beach is a coastal area in California that's abundant in sea glass, accumulated from years of garbage dumped along the coast. Decades of pounding waves wore the glass pieces down into smooth, polished pieces about the size of large pebbles. The area has actually turned into a tourist attraction, with the California Parks and Recreation Department banning the removal of the glass pieces as souvenirs.

Water is essential for all life, and clean water is a commodity with increasing demand. With this in mind, over-withdrawal and degradation of fresh water is a troubling problem. The vast majority of the human population is dependent on ground water, which can be easily contaminated by pollution on land. Even the largest reservoirs of water, the oceans, are becoming increasingly contaminated with oil and plastic waste.

## The Least You Need to Know

- The hydrologic cycle continuously moves and recycles water throughout the Earth.
- The largest available source of fresh water on Earth is groundwater. This is also one of the most easily contaminated sources.
- The biggest source of water withdrawal is irrigation, which grew markedly after the Green Revolution.
- Water pollution is most easily dealt with by not releasing it in the first place. Sewage treatment systems achieve this by harnessing bacteria to break down waste before it's released.
- The biggest source of oil pollution in the ocean is nonpoint—small amounts of dumping by individuals.

# CHAPTER 19

# Air Pollution

Chapter 17 gave a preview of some of the major risk factors when it comes to human health. While some of them are out of our control, such as bacterial disease, many of the chemical toxins are our own creation.

Pollution—contaminants that have an adverse effect on living organisms—spreads through air and water. These are the two most common routes because we all have to breathe, and we all need water.

Of the two, air pollutants tend to spread faster and farther. They're carried away by the wind, reaching destinations that may be very far away from their origin. Air pollution is a problem that we as a society have begun to tackle only recently, and like many other environmental problems, it first took some big wakeup calls.

## In This Chapter

- The most prevalent forms of contaminants in the air
- How different forms of air pollution affect living organisms
- The atmospheric reactions that cause smog, acid rain, and ozone depletion
- Technologies that reduce or eliminate air pollution

# Learning the Hard Way

Air pollution is a chronic problem anywhere there's a lot of industry and a large population. However, there have been a few incidents so terrible, with so many casualties, that they live in environmental infamy.

Each of the following incidents actually began with a normal environmental phenomenon called a temperature inversion. If you recall from Chapter 6, temperature drops as you increase in altitude (think snow-capped mountains). This is a good general rule of thumb, but it doesn't always work this way.

Places where winds are calm, such as valleys, can actually see a switch. A pocket of cool air rests at the lower elevations, while a layer of warm air lingers above. This creates a potential for some air pollution issues. Why? Remember the old rule: warm air rises, cool air sinks. What happens if you pollute into a pocket of cool air in a temperature inversion? The pollution is trapped. It doesn't go anywhere; it accumulates.

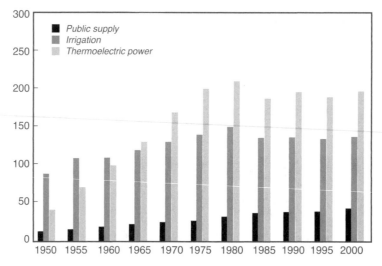

*During a temperature inversion, a layer of cooler air descends over the city. This prevents pollution from rising into the atmosphere and escaping.*

## Donora Fluoride Fog

The city of Donora, just south of Pittsburgh along the Monongahela River, found itself in a temperature inversion during Halloween of 1949.

Donora was a company town of about 14,000 people. The main employer was a zinc-smelting plant. In order to melt and process zinc or any other metal, you need a lot of heat. This heat was generated by burning coal by the trainload.

Pollution was a way of life for towns with heavy industry like this along what has been called the Rust Belt. Many of the older houses in Pittsburgh, for example, have small bathrooms right as you enter so the residents could wash off the ash and soot when they came in.

During this temperature inversion, which ended up lasting about five days, the pollution from the zinc plant built up, growing thicker and thicker. Visibility dropped to just a few feet. Driving was impossible. Streetlights were on in the middle of the day. The operators of the zinc plant refused to stop operations, knowing that doing so would come at a huge cost to their bottom line.

About half of the town's population contracted some sort of acute respiratory problem during this time. The death toll was relatively low; only about 20 people died. Nevertheless, this incident attracted a great deal of national attention.

## London Smog of 1958

London was many times more populated than Donora. It wasn't a company town; its sources of pollution were much more diverse. Most of the city's industry was very close to the residential areas, and residents all burned a type of coal called "sea coal" to heat their homes.

Sea coal is a very soft coal that was originally used to weigh down empty ships that were travelling back to England from trade routes. The coal wasn't really considered useful at the time, but after the Industrial Revolution it became much more valuable. When burned, sea coal produces quite a lot more pollution than harder forms of coal, such as that used in Donora.

The winter of 1958 was a bitterly cold one in London. Residents began burning the sea coal in greater amounts. The pollution began to accumulate, as London naturally has calm winds and is famous for sea fog as thick as pea soup.

The exact number of fatalities is unknown, but it's estimated to be over 12,000. Not long after this incident, the United Kingdom passed a law called the Clean Air Act, which among other changes actually specified an increase in the height of chimneys to deposit pollution above the natural temperature inversion.

# The Clean Air Act

The most significant parts of the United States Clean Air Act were passed in 1970. Enforcing this law was actually the first major task for the Environmental Protection Agency (EPA), which was founded the same year.

The law required the EPA to set and enforce limits for six different air pollutants. These are referred to in the law as criteria pollutants, and were considered to be the most significant risks to human and environmental health at the time. The pollutants included:

- Sulfur dioxide

- Carbon monoxide

- Particulates

- Ozone

- Nitrogen oxides

- Lead

The Clean Air Act was very successful. The effects of this law have been measured in many ways, but one of the most interesting examples is actually found within the ice cores taken from glaciers in Greenland.

Sulfur dioxide levels rose dramatically beginning around 1900. This coincides with the Industrial Revolution. Nitrogen oxides increased around the 1950s, reflecting a boom in the number of vehicles being driven. Levels of both pollutants stabilized in the 1970s.

## Sulfur Dioxide

Sulfur dioxide ($SO_2$) is a pollutant that is produced by burning coal. Remember, sulfur is one of the essential elements of life, and coal was at one point plant matter.

This pollutant is easiest to detect by its smell. Sulfur carries the odor of rotten eggs. The offensive smell is the result of sulfur dioxide as a *primary pollutant*. This is actually the least of its effects.

Sulfur dioxide reacts with water vapor in the air to produce a *secondary pollutant*: sulfuric acid. This is part of what makes up acid precipitation.

 **DEFINITION**

A **primary pollutant** is a contaminant that is released directly into the air. A **secondary pollutant** forms indirectly as a result of a chemical reaction between a primary pollutant and sunlight or water. **Acid precipitation,** which is formed when the pH of rain, snow, or fog is unnaturally lowered due to a reaction of water with sulfur dioxide, is an example of secondary pollution.

## Nitrogen Oxides

Nitrogen oxides ($NO_x$) are somewhat visible; they appear as a reddish-brown gas in the air. They're primarily a product of internal combustion engines—ones that burn products of crude oil like gasoline and diesel fuel.

Nitrogen oxides also form secondary pollutants after being released into the atmosphere. First, they can combine with water vapor to produce nitric acid. Second, they react with sunlight to form the much more visible smog. This should make sense—after all, what cities do we most associate with smog? Los Angeles and Beijing, places with a lot of cars on the road producing a lot of exhaust.

## Carbon Monoxide

Carbon monoxide (CO) is another pollutant closely tied to automobile exhaust. This molecule can be created anytime something is burned, but is produced in greatest amounts by fossil fuel–consuming engines.

Carbon monoxide is the most insidious of the air pollutants because it's the most dangerous; yet you can't see it or smell it at all. When inhaled, carbon monoxide binds to the hemoglobin protein in red blood cells. This interferes with their ability to absorb and transport oxygen.

Your entire body needs oxygen, but you first feel the effects of its deprivation in your brain. The initial symptoms are dizziness and sleepiness. Carbon monoxide is the main reason why you are never supposed to turn on your car in the garage without first opening the door.

## Particulates

Particulates (PM) are exactly what they sound like: particles of dust, ash, soot, or any other visible material. When you think of the smoke pouring out of a chimney, what you're actually seeing is particulate pollution.

Particulates are separated into two categories based on the size of the particle.

- Particles less than 2.5 microns in diameter are labeled PM2.5.
- Particles up to 10 microns in diameter are labeled PM10.

All particulate pollution has the effect of diminishing visibility. However, the smaller particles are the ones that are most likely to be inhaled and embed themselves into respiratory tissue.

# Lead

Lead (Pb) might seem misplaced in this law. As we learned in Chapter 17, it's certainly a dangerous neurotoxin. However, it's a solid metal. How or why would it be airborne?

Small particles of lead can attach themselves to particulates, and are inhaled along with the particulates. This pollutant was added to the Clean Air Act of 1970 because a great deal of concern was being raised about the effects of lead poisoning on the IQ levels of children.

 **A LOOK BACK**

> A compound containing lead was used as an additive for gasoline during the first several decades following the invention of the internal combustion engine. Concerns about the neurotoxic effects of lead pollution resulted in this gasoline being phased out by 1995. Currently, only a handful of countries still have leaded gasoline even available, and modern engines no longer require the additive.

Today, airborne lead pollution is much less prevalent. The single biggest source is burning fossil fuels, particularly coal.

# Ground-Level Ozone

The Clean Air Act was amended in 1990. Several additional standards were added for pollutants that were not addressed by the initial law. One of those is ground-level ozone ($O_3$).

You've probably heard of this gas before in reference to the ozone layer. That's something much higher in the atmosphere and not related to this problem. Ground-level ozone is highly toxic to all living organisms.

Ozone is not released directly. It's a secondary pollutant, the result of a mixture of nitrogen oxides and volatile organic compounds.

# Volatile Organic Compounds

Volatile organic compounds (VOCs) have two important properties. First, they often begin as liquids but evaporate easily. This is where the "volatile" part comes from. Second, they're complex organic (carbon-containing) chemicals.

One common example of a volatile organic compound is gasoline. If you've ever accidentally spilled gasoline while filling your car, you might have noticed that it evaporates pretty quickly. This (along with the danger of fire) is the reason why you aren't supposed to top off your tank when filling up.

The other big source of VOC pollution is solvents. Solvents are chemicals that dissolve other things. One example you may have experienced is paint thinner. The fumes given off by this solvent are intense, which is why you have to have good ventilation when painting.

# Secondary Effects of Air Pollution

The effects of primary pollutants, at least, are pretty straightforward. Each of them causes direct, specific effects on the health of humans or other organisms. Air pollution becomes quite a bit more complicated when we start to look at all the secondary pollutants that form later.

Once released, pollutants will interact with other compounds already in the air. Sunlight is present and often serves as a catalyst behind these reactions.

Each of the secondary effects we'll discuss in this section has its roots in one specific substance that humans released. Remember that each of these is preventable, assuming that the original pollution is not produced in the first place.

## Smog

Smog has become an all-encompassing word to describe any situation where the air takes on a hazy, polluted appearance. There are actually a few different types and sources of smog.

The smog that affected Donora and London before the passage of the Clean Air Acts was industrial smog. This type of smog was mostly particulate pollution (mixed in with sulfur dioxide and whatever other invisible compounds were released) combined with some naturally occurring fog.

 **A LOOK BACK**

Industrial smog grew rapidly during the early decades of the Industrial Revolution. In his book *Hard Times,* Charles Dickens wrote this description of the industrializing countryside:

"Coketown lay shrouded in a haze of its own, which appeared impervious to the sun's rays. You only knew the town was there because there could be no such sulky blotch upon the prospect without a town."

Photochemical smog, as the name suggests, is the product of light reacting with specific chemicals. For the most part, these chemicals are nitrogen oxides.

Photochemical smog is highly associated with traffic. Densely populated cities will regularly experience smog especially during the morning commute. The number of cars on the road is great, the temperature is starting to rise, and the sun is out. The smog that's produced as a result of all this exhaust often lasts until the evening rush hour.

## Acid Precipitation

Rainwater is normally slightly acidic, with a pH of about 6. This is mostly due to the dissolving of some carbon dioxide in the air in the water vapor, creating carbonic acid. With the onset of the Industrial Revolution, many more acids began to form in the atmosphere, creating a growing problem of acid precipitation.

The main culprit is sulfur dioxide. Other pollutants, such as nitrogen oxide, can also form acids, but sulfur dioxide is by far the biggest contributor. Do you remember what the biggest source of sulfur dioxide is? Coal! What facilities burn lots of coal? Power plants and metal smelters.

There are coal-fired power plants all over the United States, but the largest emitters are located along the east coast and the Midwest. Remember, most of the steel and zinc plants in the United States are located in the Rust Belt between Illinois and Pennsylvania.

Field PH

5.3  5.2-5.3  5.1-5.2  5.0-5.1  4.9-5.0  4.8-4.9  4.7-4.8  4.6-4.7  4.5-4.6  4.4-4.5  4.3-4.4  <4.3

*The mid-Atlantic and New England states tend to have the lowest soil pH due to the plume of sulfur dioxide emanating from the Rust Belt.*

Why is it damaging to lower the pH of precipitation? First, it interferes with the ability of plants to absorb nutrients from the soil. Trees and other plants have evolved to maximize the efficiency of their roots with a certain pH. We've changed that pH in a relatively short period of time. Aquatic organisms—fish and amphibians—are also highly impacted by this problem.

As humans, we aren't in any danger of suffering acid burns when we walk outside. The pH of acid rain at its worst may approach that of vinegar. Our skin can easily withstand that. However, many of our buildings and other structures, especially ones made of materials like limestone and marble, will gradually dissolve and break down when exposed to acid precipitation.

## A Hole in the Ozone Layer

The secondary effect that gave us the biggest surprise was the depletion of the *ozone layer.* This is a layer of ozone ($O_3$) up in the stratosphere, the layer of the atmosphere we only visit when travelling by airplane.

About 30 years ago, scientists observed that a huge hole in this layer was forming over the continent of Antarctica. This was alarming because we need the ozone layer to filter out some of the ultraviolet radiation from the sun. Ultraviolet radiation is a mutagen; in high amounts, it's damaging to living tissue.

Why was this hole forming over Antarctica? Why did it grow the most during the months of September and October every year? The answer, we discovered, was in *ozone-depleting substances* we were emitting in pollution called *chlorofluorocarbons (CFCs).*

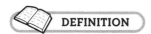 **DEFINITION**

> The **ozone layer** is a part of the stratosphere that contains high amounts of ozone ($O_3$). This molecule absorbs some of the ultraviolet light that reaches the Earth from the sun. **Ozone-depleting substances** are airborne compounds like **chlorofluorocarbons (CFCs)** that cause the degradation of ozone into oxygen ($O_2$).

This large, chlorine-containing molecule was ascending into the stratosphere and causing ozone to break down, destroying its ability to absorb ultraviolet light. It was doing so over Antarctica because the swirling of winds over the Earth was concentrating the CFCs there. The worst was during Antarctica's spring (September and October), when some of the snow and ice would melt, releasing CFCs that had been trapped the rest of the year.

Fortunately, this was one environmental problem we were able to act upon. CFCs were primarily used in aerosol cans and as refrigerants. Cost-effective substitutes were found, and in 1989 an international treaty called the Montreal Protocol went into effect. CFC pollution dropped rapidly, and the ozone layer is gradually starting to recover.

# Keeping the Air Clean

As with any mess or disaster, the easiest and cheapest solution is to prevent it from happening in the first place. The cost of preventing just about any problem ends up being much cheaper in the long run, but our society isn't very good about using that kind of foresight.

Now the primary and secondary effects of air pollution are understood. We've experienced a few major disasters, but lessons have been learned, and regulations have been created to limit pollution. Let's take a look at some of the technologies and methods now in place.

## Filtration and Precipitation

In many industries, air pollution is unavoidable. If we choose to burn coal for electricity or burn gasoline to power our cars, sulfur dioxide, nitrogen oxides, and other pollutants will be produced.

Some smokestacks now are equipped with scrubbers or porous filters that trap and remove particulates. Others have electrostatic precipitators, which are electrically charged plates that serve as magnets to attract particles of ash and soot.

Modern cars are now all equipped with catalytic converters. These devices contain two chemical compounds (catalysts) that promote reactions that neutralize nitrogen oxides, volatile organic compounds, and other exhaust emissions. Between catalytic converters, better engine design, and greater efficiency, cars produced today are over 90 percent cleaner than cars from the 1970s.

## Better Fuels

As much as technology to reduce emissions has improved over the last few decades, the fact remains that certain pollutants could be eliminated entirely if the fuels that produced them were simply not used.

Recall the events that led to the London smog disaster of 1952. People were burning large amounts of sea coal, one of the dirtiest fuels known. What do you use as a source of heat in your home today? Chances are, especially if you live in the United States, you use natural gas.

Natural gas is often labeled the "clean" fossil fuel because it produces far fewer emissions than either coal or the products of crude oil. In fact, the only pollutant produced from burning methane is carbon dioxide. While carbon dioxide pollution certainly has its own set of problems (see the discussion of global warming in Chapters 20 and 21), it doesn't directly affect our health. Furthermore, if we switch electricity generation over to renewable sources like solar, wind, and hydro, the pollution produced from those is zero.

With the onset of the Industrial Revolution, air pollution became a growing problem. Only the passage of the Clean Air Act finally stemmed the increase of these pollutants. Now we're working to find new fuels and technologies to feed the growing energy demands of our society without repeating the great air pollution tragedies of the past.

## The Least You Need to Know

- Primary air pollutants are released directly into the air, while secondary pollutants form as a result of chemical reactions in the atmosphere.
- Acid precipitation is largely the result of sulfur dioxide, a product of burning coal.
- Smog is produced when nitrogen oxides from car exhaust react with sunlight.
- The Clean Air Act, passed in 1970 and amended in 1990, sets standards for how much air pollution any given plant can produce.
- Air pollution can be decreased by implementing technologies that remove it from emissions or choosing a cleaner source of energy.

# The Greenhouse Effect

In Chapter 19, we covered most of the major air pollutants. One emission we didn't mention that's found in just about any kind of combustion (burning) reaction is carbon dioxide.

For a long time, carbon dioxide wasn't really considered a dangerous air pollutant. Yes, it's toxic to humans at high enough concentrations, but nowhere near the same level as carbon monoxide or sulfur dioxide.

As a comparison, consider the lethal concentrations (about 30 minutes of exposure) of these common emissions from burning fossil fuels.

## In This Chapter

- The different layers of the atmosphere
- The greenhouse effect, and the gases that cause it
- Climatic factors that influenced the major ice ages
- How climate shifts have influenced pre-industrial human society

| Gas | Formula | Lethal Concentration |
| --- | --- | --- |
| Carbon Monoxide | CO | 0.2% |
| Carbon Dioxide | $CO_2$ | 10% |
| Sulfur Dioxide | $SO_2$ | 0.3% |

So when the Clean Air Act was signed in 1972 and given to the EPA to enforce, carbon dioxide was not one of the criteria pollutants. When a series of amendments to the law were passed in 1990, carbon dioxide still was not included on the list.

Carbon dioxide may not be a huge danger to human health in the short term, but our production of it may have some serious consequences for all life on Earth in the near future.

# What Is the Atmosphere?

Carbon dioxide is actually only one of several gases that make up our atmosphere. It's been part of that mixture for as long as life has existed on Earth.

The Earth's atmosphere is a very thin layer of gases that's exceptionally important to supporting life. If the Earth were an apple, the atmosphere would be even thinner than the apple's skin. The atmosphere contains the oxygen we breathe, nitrogen used by bacteria and plants, water vapor, the ozone layer, and a thick layer of gases that serves as a blanket to hold in some of the heat the Earth absorbs from the sun every day. Carbon dioxide is part of that blanket.

## The Outer Reaches

The uppermost part of our atmosphere is the exosphere (*exo* means "outer"). The exosphere is very thin and serves as the transition between our atmosphere and the relative emptiness of space.

The thermosphere (*thermo* means "heat") is a very warm layer about 90km to 1,000km above the Earth's surface. Temperatures in this layer fluctuate wildly, increasing 200°C (360°F) during the daytime. Space shuttles and the International Space Station orbit Earth in this layer.

The mesosphere (*meso* means "middle") is about 50km to 85km above the Earth's surface. Not as much is known about this layer, because it's too high for planes and weather balloons to reach, but below the orbit of satellites. What is known is that meteorites burn up in this layer.

In the overall design of the Earth's atmosphere, these upper layers are important; but as humans, we don't interact directly with them very much. Most of what we see and breathe is in the lower two layers.

## The Innermost Layers

The stratosphere is the second-closest layer to the Earth's surface, beginning at about 16km up. This layer is very calm and cloud-free, a big reason why commercial airplanes travel up to this altitude. The ozone layer, which protects us from ultraviolet radiation, exists in this layer.

The last layer is the troposphere (*tropo* means "change"). This layer contains the air that all living things breathe, clouds, precipitation, and wind currents. In other words, it contains our weather. The troposphere also contains the gases that produce what we call the greenhouse effect.

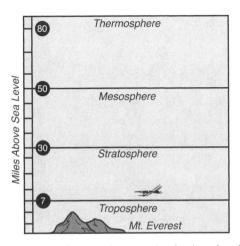

*The bottom layer of the atmosphere, the troposphere, contains the air we breathe and the water cycle. The upper layers, while important, do not have as direct an influence on climate.*

# Staying Warm

The greenhouse effect is one of the most important functions the atmosphere provides to support life. Consider the thermosphere. Temperatures in this layer can fluctuate several hundred degrees, depending on the time of day and how much energy the sun is producing at that moment. Most living things wouldn't be able to survive these fluctuations.

Even in the most dynamic of biomes, temperatures stay within a certain habitable range throughout the course of the day or year. This is one of the aspects that make the Earth so welcoming to life compared with other planets. These stable levels are primarily due to influences of the lower atmosphere.

# Energy from the Sun

As we covered in Chapter 8, energy flows through an ecosystem. Looking at the big picture of the entire biosphere, energy flows in from the sun during the day, then exits into space during the night. Life wouldn't exist without this continuous flow of energy from the sun.

Energy flows toward the Earth in a pattern of waves, much like the waves of the ocean. Each form of energy has a different wavelength.

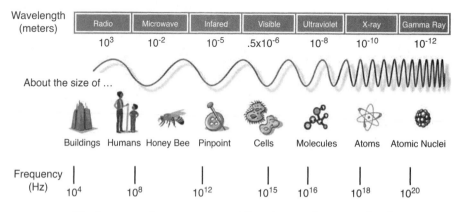

*The electromagnetic spectrum, arranged from shortest to longest wavelength.*

The wavelengths of energy produced by the sun that are most relevant to us are ultraviolet, visible, and infrared. Ultraviolet light, due to its shorter wavelength, can penetrate living tissue. This is why overexposure to the sun causes burns, and over the long run, an increase in the risk of skin cancer. Humans cannot detect the wavelength of ultraviolet light or anything shorter without specialized equipment.

Visible light is the range of wavelengths our eyes can detect. Each color of visible light represents a different wavelength. Photosynthetic organisms like plants harness certain wavelengths of visible light to convert low-energy carbon dioxide into high-energy glucose.

**CASE STUDY**

A common misconception about plants is that they perform photosynthesis with green light, hence their color. In reality, green plants actually absorb just about every wavelength of light except green! This color is reflected, and we see it. This is the reason why snow is white, as it reflects all colors, while soil is black because it absorbs all colors.

Infrared light, with a wavelength longer than is visible, isn't detectable to our eyes. However, we can feel it! If you stand outside in the sun and feel the warmth, that's infrared energy. Infrared behaves like visible light in that it can be reflected by some surfaces and absorbed by others.

## The Greenhouse Effect

Of all the infrared energy that hits the Earth's atmosphere, only about half reaches the surface. The rest is either reflected back into space or absorbed by the upper layers of the atmosphere.

What happens to the infrared energy that does reach the surface depends on what it strikes. Clean, white snow will reflect most of that energy back into the atmosphere. Dark soil will absorb most of it.

All that reflected energy, in addition to any heat released by living organisms on the surface (including us), would normally just escape right back into space. However, certain components of the atmosphere actually retain some of this lost heat, then re-radiate it. The heat energy then continues to circulate throughout the troposphere.

This phenomenon is called the greenhouse effect because it works in a similar way to a greenhouse constructed out of glass or other transparent material. Energy from the sun is able to penetrate through the clear glass and warm the inside. As the heat rises, it's trapped by the solid glass roof. The greenhouse continues to warm, reaching a higher temperature than the outside environment. The analogy isn't perfect, as the greenhouse gases actually absorb and re-emit the heat, unlike the glass roof, which simply reflects it.

The gases that contribute to this are called greenhouse gases. They include those listed in the following table.

| Gas | Formula | Contribution to Greenhouse Effect |
| --- | --- | --- |
| Water vapor | $H_2O$ | 36% to 72% |
| Carbon dioxide | $CO_2$ | 9% to 26% |
| Methane | $CH_4$ | 4% to 9% |
| Ozone | $O_3$ | 3% to -7% |

*Data from* Earth's Annual Global Mean Energy Budget, *by Kiehl and Trenberth.*

Most of the focus in terms of greenhouse gases is placed on carbon dioxide and methane. Why? Water vapor is going to be naturally present in the atmosphere. It's part of the water cycle and not really within our control. Carbon dioxide, however, is something we as a society produce a great deal of. Methane, while not as prevalent in the atmosphere, is actually a much more potent absorber and emitter of infrared energy.

## Heat and Climate

You've undoubtedly heard of global warming before. This term is sometimes used to describe changes occurring on the Earth as a result of the greenhouse effect. This isn't really accurate, because the Earth isn't just getting warmer. The climate itself is changing.

The different climates across the biosphere—precipitation, average temperature, humidity, and so on—are driven by heat. The entire water cycle, for that matter, occurs as a result of heat.

Water evaporates from bodies of water or transpires from plants as a result of heat. As the water vapor ascends into the atmosphere, it cools, condenses, and falls back down due to a lack of heat. Do you remember the two biggest variables that determine a biome? They're temperature and precipitation. Temperature is a measurement of heat, and precipitation occurs as a result of heat.

With this in mind, the importance of average temperature across the Earth should come into focus. A shift of a few degrees cooler or warmer can have some pretty monumental effects on our climate. We know this because it has happened before.

# Climate Change Is a Natural Process

We already know that the overall climate of the Earth has changed dramatically from age to age. As we covered in Chapter 2, there have been five known ice ages, each with its own glacial periods and interglacial periods.

 **A LOOK BACK**

Snowball Earth is a hypothesis that the Earth's surface was nearly entirely frozen at one point in time. Evidence for this hypothesis centers around sediments found at the bottom of subtropical oceans that normally would only be found toward the poles. The cause of such a global freeze is tied to carbon dioxide levels. One explanation suggests that the continents had all drifted near the equator, resulting in more precipitation falling on land. Some carbon dioxide dissolves in rainwater, and when it falls on land it can react with rocks to form carbonate, effectively locking it out of the atmosphere.

The known ice ages, and their approximate time periods, are listed in the following table.

| Name | Period (Millions of Years Ago) |
|---|---|
| Huronian | 2400 to 2100 |
| Cryogenian | 800 to 635 |
| Andean-Saharan | 450 to 420 |
| Karoo | 360 to 260 |
| Quaternary | 2.58 to Present |

With the exception of only a recent sliver of the Quaternary Age, humans had no influence on the Earth during these ice ages. They occurred due to natural causes.

Let's take a look at the Karoo Ice Age. The beginning of this age coincides with the Carboniferous period, which we mentioned in Chapter 14 as the time in which most of our coal was formed.

This ice age was caused by plants. Specifically, it was caused by an incredible abundance of plant life and a lack of any decomposer able to break down lignin in wood. Remember, trees perform photosynthesis, which sucks carbon dioxide out of the air and converts it into glucose that the plant uses as a building block to make all its other tissues, including lignin.

Carbon dioxide levels dropped during this age. Summer temperatures cooled so much that the previous winter's ice accumulations weren't fully melted. Glaciers grew and extended over much of the Earth's surface.

Eventually, the Earth's climate cooled enough to slow down plant growth. Elevated oxygen levels resulted in massive fires. Eventually, bacteria and fungi evolved a way to break down lignin. Carbon dioxide was returned to the atmosphere, and the ice age ended.

## The Last Ice Age

What about the Quaternary Ice Age, the one we're now in? Unfortunately, that answer isn't as straightforward as with the Karoo Ice Age, but there are a few suspects.

The orbit and tilt of the Earth can change slightly. The tilt, or angle of the Earth's axis, can vary a few degrees, causing a greater difference between the seasons.

> **CASE STUDY**
>
> Nearly everyone is familiar with their zodiac sign. Each sign has a range of dates when the sun crosses a constellation as it rises. If you were born between July 22nd and August 22nd, you would expect the sun to cross Leo, the sign of the lion, during this period. However, it probably won't. The Earth has wobbled slightly on its axis since the zodiac positions were determined over 2,000 years ago. Instead, the sun crosses Leo from August 16th to September 15th! The zodiac signs as we know them are nearly a month off!

Studies of ice cores in Greenland have shown a decrease in carbon dioxide levels leading up to this ice age. However, the cause of this increase and its actual impact on the onset of global cooling are unknown.

## The Medieval Warming Period

So far, this is all ancient history. These huge shifts in climate occurred well before the earliest *Homo sapiens* appeared. There have been some smaller changes in recorded history, both warming and cooling.

The Medieval Warming Period occurred sometime between 900 and 1300 C.E. During this time, summers were longer and winters were milder throughout Europe. This led to some abundant harvests, particularly in Western Europe. The Vikings were able to travel through ice-free seas and colonize Greenland and Labrador during this time. Meanwhile, much of South America experienced a severe drought. This is suspected to be a contributing factor to the downfall of the Mayan Empire.

The exact causes of this cooling period are not known, but there may have been a change in the North Atlantic Oscillation—westerly winds that blow across the Atlantic and bring moist air into Europe. Whatever the causes were, the effects of the Medieval Warming Period were pretty minor and temporary in comparison to past ice ages. Yet these changes had a big impact on our society.

## The Little Ice Age

The Little Ice Age was a period of cooling that followed the Medieval Warming Period, lasting from about 1300 to 1800 C.E. Europe and North America experienced colder winters. There are records of winter festivals in Britain during this time, with ice skating along the rivers.

The population of Iceland fell by half, and the Viking colonies in Greenland vanished. Agricultural yields, which had been so plentiful during the Medieval Warming Period, dropped sharply, leading to a series of famines.

Again, the exact cause of this cooling is not known. Volcanic activity may have played a role. Volcanic eruptions release a lot of particulates and sulfur dioxide into the stratosphere, where they block out some of the incoming sunlight.

There are two important lessons to be taken from the Medieval Warming Period and Little Ice Age. First, the Earth's climate system is incredibly complex and can be altered by variables that are completely out of our control. Variables like volcanic eruptions, changes in the Earth's tilt, and overall wind currents all play a role. It would be a mistake, however, to assume that every variable is in the hands of nature and not man. Remember, a good deal of the greenhouse effect is caused by carbon dioxide and methane gas, two substances our society is known to produce and release in massive quantities.

## The Least You Need to Know

- The troposphere is the layer of the atmosphere that we breathe, that contains weather, and that most of our pollution escapes into.
- Greenhouse gases contribute to the ability of the atmosphere to retain and emit infrared (heat) energy that escapes from the surface.
- As seen throughout the ice ages, the Earth's climate has shifted dramatically without any influence from man.
- Naturally existing factors that influence climate include volcanic eruption, shifts in the Earth's tilt, and photosynthetic activity of plants and algae.

# Anthropogenic Climate Change

In Chapter 20, we looked at some of the naturally occurring variables that have influenced the Earth's climate over the last few billion years. You may have noticed the absence of any mention of the human effects—impacts our actions have had, especially since the Industrial Revolution.

In this chapter, we're covering anthropogenic climate change. Anthropogenic simply means "man-made" or "human-generated." Notice that we aren't calling this global warming, because that's oversimplifying the problem.

An increase in the overall heat levels of the Earth doesn't just make temperatures warmer. If that were the case, everyone living in the northern United States and Canada might welcome a few extra weeks of summer. No, we're talking about a change in climate: rainfall, evaporation, ocean currents, wind currents, and everything in between.

## In This Chapter

* The observed changes on Earth from global warming
* Research linking global climate change to human activities
* Prevention strategies that help keep greenhouse gases from being released
* Cleanup strategies that remove and lock away carbon dioxide already in the atmosphere

# How Do We Know?

Global climate change is often painted as a debate. It may be given a label like "scientific uncertainty." The fact is, there isn't as much of a debate about global climate change as you might think.

The Earth is warming. That part isn't really in question, and plenty of data is available to support that, which we'll look at in the next section. The increase in heat is beginning to impact our climate.

That part is also generally accepted, at least by scientists. What about the "anthropogenic" part? How do we know it's human-caused? Well, it's pretty hard to prove that any of these impacts are caused by humans. To be honest, it's often difficult for science to prove anything. What we do know is that there aren't any other peer-reviewed explanations backed up by data. This section is all about what we do know, and how we know it.

## The IPCC United Nations Report

One of the biggest and most highly regarded sources of data and discussion about the causes and effects of global climate change is the Intergovernmental Panel on Climate Change (IPCC). This is a scientific body founded in 1988 by the United Nations.

The way the IPCC works is thousands of scientists contribute reports on a voluntary basis. The reports are reviewed by representatives from the UN member governments, who then write a summary that's made available to the public.

 **CASE STUDY**

According to the IPCC Fourth Assessment Report (2007), "Warming of the climate system is unequivocal." ... "Most of the observed increase in global average temperatures since the mid-20th century is very likely due to the observed increase in anthropogenic greenhouse gas concentrations."

The IPCC doesn't do any of its own research. Rather, it's a confluence of research from other organizations and universities all over the world. The IPCC (along with Al Gore) received the 2007 Nobel Peace Prize for these efforts. The IPCC is very well respected and considered an objective body. All of the data we'll examine in this chapter comes from the IPCC.

## Global Average Temperature, Sea Level, and Snow Cover

The IPCC puts out a new report about every five or six years. Its fourth and most recent report came out in 2007. The summary of this report included three graphs, each representing recorded changes in our climate since the late 1800s, when consistent written records of temperature and precipitation began.

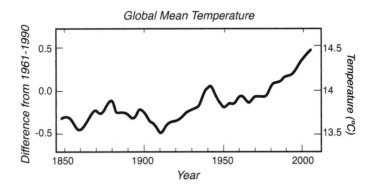

The first graph is of global average temperature. The data shows an increase of about 0.74°C globally over the last century. Eleven of the most recent years included in the graph (1995 to 2006) rank among the 12 warmest years in the instrumental record (since 1880).

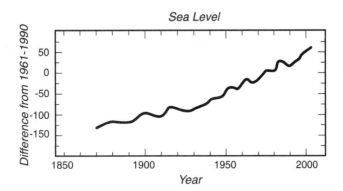

The second graph covers ocean levels. According to the report, sea levels rose an average of 1.8mm per year from 1961 to 2003. The average increase between 1993 to 2003 was 3.1mm per year. Part of this is due to the natural expansion of water when it warms. The rest is explained by the third graph.

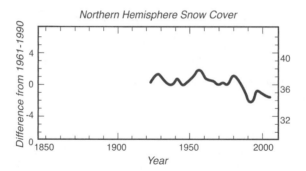

The last graph looks at snow cover in the Northern Hemisphere. Only the Northern Hemisphere was included because the majority of the world's land mass is north of the equator, and so far, sea ice in Antarctica has not shown any changing trends. The data does show that mountain glacier size and snow cover (including Greenland and the Arctic) are diminishing.

## Measurements from the Past

This data is pretty compelling, but it only paints part of the picture. After all, the Earth is billions of years old, and we only have written records going back about 130 years. Short of someone inventing a time machine, how do we find out what global temperatures and carbon dioxide levels were like thousands or hundreds of thousands of years ago?

**A LOOK BACK**

In 1958, Charles David Keeling began taking measurements of carbon dioxide in the atmosphere in Mauna Loa, Hawaii. This location was far enough away from major cities or forests to be considered a good, unbiased sample. The atmospheric carbon dioxide was found in greater concentrations in winter and lower in summer. The reason has to do with photosynthesis: there are more plants absorbing carbon dioxide during the summer, causing that temporary decrease.

The answer is proxies. Proxies are ways of measuring something indirectly. Scientists use proxies all the time. Carbon-14 dating is a way to indirectly measure age. Tree rings can be studied to see which years the surrounding ecosystem was experiencing a drought.

For climate scientists, the proxy of choice is the ice core. As ice builds up in the polar regions of Earth, tiny bubbles of atmosphere are locked inside. When the ice is buried, the bubbles of atmosphere become isolated from the outside environment. The deeper the ice core, the older the bubbles of atmosphere present.

These tiny samples of ancient atmospheres can be analyzed to determine two very important pieces of data. First, the percentage of greenhouse gases like carbon dioxide in the sample. Second, the ratios of two different isotopes of water can be measured to infer the temperature conditions of the Earth. For example, during colder periods of the Earth's history, a lighter isotope of water is more likely to evaporate from the oceans.

## The Vostok Graph

During the early 2000s, the longest ice core ever drilled was taken at the Vostok research station in Antarctica. The core was long enough to date back 420,000 years, revealing a total of four ice ages. Data from this ice core was used to construct a graph comparing carbon dioxide levels and global temperature.

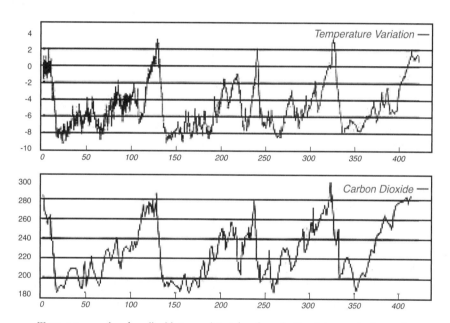

*Temperature and carbon dioxide proxy data taken from the Vostok ice core in Antarctica.*

The visual similarities between the two variables are striking. There certainly seems to be some correlation between carbon dioxide and global average temperature levels. An even longer ice core was drilled in Antarctica a few years later, and that data produced a similar pattern.

What conclusions can we draw from this data? First, carbon dioxide levels have increased markedly in the last century. And temperature levels have also increased. Notice that the Earth has experienced levels similar to this before—about 140,000 years ago, before humans walked the Earth. Truly, our society is entering uncharted climate territory.

**CASE STUDY**

In the documentary *An Inconvenient Truth,* Al Gore showed his audience the Vostok carbon dioxide and average temperature graph. He also projected alarmingly high carbon dioxide levels in the future. He concluded by saying, "Ultimately, this is really not a political issue so much as a moral issue. If we allow that to happen, it'd be deeply unethical."

## Is That Our Carbon?

The Vostok graph looks pretty alarming, at least from the human perspective. Remember, though, that Earth's climate has gone through some pretty monumental changes, and these all took place well before the earliest humans entered the picture. How do we know that the changes taking place now, spurred by the rise in carbon dioxide, are really due to our influence? Can humans really be producing enough $CO_2$ to affect the climate of the entire globe?

One of the proxies scientists use to analyze ancient atmospheres is to look at the ratio of different isotopes of oxygen atoms in water. We can do the same thing to carbon atoms in carbon dioxide.

There are three isotopes of carbon: Carbon-12, Carbon-13, and Carbon-14. Each atom is identical, except for the number of neutrons. Living organisms are made of a mixture of all three isotopes. Fossil fuels, which were originally living tissue, don't contain any carbon-14, as it's unstable and breaks down over time.

The isotope "fingerprint" of the atmosphere is changing. Plants have a lower ratio of Carbon-13 to Carbon-12, and the atmosphere is more and more resembling this ratio. The explanation that most scientists accept is that this change can only be the result of the burning of massive amounts of fossil fuels.

There is a lot of data and big ideas to consider here, but it boils down to one key relationship. If you agree that carbon dioxide levels are rising and accept the likelihood that carbon dioxide is behind the recent increase in global temperature, then you must also accept that most of the increased carbon dioxide levels are anthropogenic.

# Consequences of Climate Change

When students learn about climate change, one of the first reactions is that humans are "destroying the Earth." That isn't really true. The Earth will continue on through its climatic cycles, just as it has for billions of years. The more apt concern is for our society. What degree and types of changes are we in store for? What changes are we already experiencing?

## Changes in Biomes

The primary concern we should have regards our system of agriculture. There are certain areas of the world that are depended upon to produce huge yields of food.

The United States Department of Agriculture (USDA) has defined different hardiness zones throughout the country. These zones reflect the length of the growing season in each area. They help growers determine what plants will be successful in any given zone.

The hardiness zones have changed since 1990. For example, about two thirds of Illinois used to be in Zone 5. Now, most of the state is in Zone 6. The first autumn frost is coming later, and the last spring frost is coming earlier.

These kinds of changes are occurring all over the globe. The range that certain species of insects, animals, and plants can inhabit is changing. Animals in the arctic, like the polar bear, are losing their habitat at an alarming rate.

**CASE STUDY**

Champagne, by definition, is sparkling wine that must come from grapes grown in the Champagne region of France. The average temperature in the region has increased by more than 2 degrees since the 1950s. According to *The Champagne Guide 2011,* this increases the risk of serious diseases and lengthens the season that certain insect pests can infect the grapes. The actual chemical nature of the champagne has also changed, becoming less acidic in recent years.

## Changes in the Water Cycle

Growing seasons aren't the only variable influencing food production. Remember from Chapter 11 that droughts have historically caused over half of all famines.

When temperatures are warmer, evaporation and transpiration occur more quickly. This puts more moisture into the atmosphere, which will eventually fall back down as rain or snow.

The end result is more extreme droughts and floods. What may have been a moderate disaster without the influences of global warming intensifies due to changes in the water cycle. Projections for the future predict a doubling, tripling, or even quadrupling of the likelihood of a summer drought, depending on the area of the world. Other regions are expected to have anywhere from 30 percent to 200 percent wetter flash flood events.

## Changes in Ocean Storms

A record-setting year for hurricanes occurred in 2005. The worst of all the storms that year was Hurricane Katrina, which flooded and destroyed much of New Orleans. Not coincidentally, when the documentary *An Inconvenient Truth* was released in 2006, promotional posters showed a smokestack emitting pollution that took the shape of a hurricane.

 **CASE STUDY**

Following the summer of 2005, Hurricane Katrina became a centerpiece in the global warming debate. The storm actually crossed Florida as a category 1, but then intensified rapidly over the warm waters of the Gulf of Mexico. So, was it caused by global warming? The answer is not really. The storm didn't form because the oceans were warmer. It may have become slightly more powerful, but it probably still would have formed. There are plenty of other ecological factors to blame. Parts of New Orleans lie below sea level, and the natural wetland buffers along the coast have shrunk as they've been built upon. These factors contributed much more directly to the devastation.

To understand the relationship between global warming and hurricanes, you need to understand that hurricanes grow fastest over warm ocean waters.

Hurricanes, like many other meteorological phenomena, are complex and cannot be predicted on the basis of just one variable. However, with the warming of the oceans, hurricanes are expected to become more intense. According to the National Oceanic and Atmospheric Administration (NOAA), the increase could be up to 10 percent. There's no known correlation between global warming and hurricane frequency.

## Changes in Ocean Water

Rising temperature is not the only change occurring in the ocean. Part of the carbon cycle is a movement of carbon dioxide between the ocean and the atmosphere. When carbon dioxide levels increase in the atmosphere (which they have), the rate at which they're absorbed into the ocean also increases.

When carbon dioxide dissolves in water, some of it is converted into carbonic acid. This addition of acid to the water lowers its pH. This is a huge problem, because many aquatic organisms are highly sensitive to changes in pH. Many shellfish and coral, for example, use compounds to produce their shells that dissolve when exposed to lower water pH.

About half of the carbon dioxide we've emitted since the Industrial Revolution has ended up in the ocean. The pH of the ocean has dropped by about 0.1. That doesn't sound like much, but remember, each pH level represents a factor of 10.

## Changes to the Coasts

As the oceans warm and become more acidic, they've also grown in size. Since 1950, global average sea level has risen between 1.5mm and 2.0mm per year. This doesn't sound like a lot, but over 60 years that works out to 9cm, or about 3.5 inches total.

Two factors are behind this rise in sea level. First, water expands as it warms. Second, the glaciers and ice sheets are melting.

How does this affect us? Make a mental list of the most populated cities you know, all around the world. How many of those are in coastal areas? Miami, New York City, Shanghai, and Mumbai are just a few. Unless the expansion of the ocean is somehow halted, some pretty big feats of engineering are going to be needed to keep these cities above water.

 **CASE STUDY**

The Maldives is a small country made up of 1,200 islands and atolls beginning just south of India and Sri Lanka. It has the unfortunate distinction of having the lowest altitude of any country in the world—the highest place barely reaches 5 feet above sea level. As the ocean levels rise, more and more of the country is eroded away. The hundreds of thousands of people who call the Maldives home may someday become the first-ever climate refugees.

## Changes to the Water Supply

About 99 percent of the world's sheet ice is located in Antarctica and Greenland. The remainder is found among the great mountain ranges of the world. These glaciers are also rapidly receding.

This is a big concern for us, because many of the great rivers of the world are fed by melting glaciers. The Himalayan glaciers, in particular, are the source of the Yellow, Yangtze, and Ganges rivers, among others. These rivers provide water and irrigation for the massive populations of China and India. Literally billions of people depend on them!

Unfortunately, in the 2007 IPCC report, an erroneous statement was included that predicted the demise of these glaciers by 2035. This statement wasn't verified and was later retracted. While errors like this are part of the scientific process, it casts some doubt in the public's eye of a problem that already seems overwhelming in scale. The glaciers very well may disappear at some point in the future, but that date is impossible to predict.

# Solutions and Prevention

In the first chapter of this book, we spent an entire section comparing the different types of environmental ethics. If you remember, most people fit into two categories: ecocentrists, who place the environment first, and anthropocentrists, who put human needs first.

Global climate change is an unusual problem in its scope. It transcends the debate between these two sides. A drastic and rapid (at least on the scale of geologic time) change in climate is bad for human society, bad for the ecosystems as we know them, and bad for the other organisms that inhabit the Earth.

If global climate change is really the one issue to "rule them all," why has fixing it been such a source of consternation? Sweeping legislation was passed in the early 1970s to deal with air pollution and water pollution. Land preservation has been ongoing since the time of Teddy Roosevelt. An international agreement was signed to end CFC emissions. Yet here we are in the twenty-first century, unable to curtail our production of greenhouse gases, especially carbon dioxide.

Let's revisit one last concept from Chapter 1: "The Tragedy of the Commons" essay. People tend to care for or use a resource differently when it belongs to them versus when it is a commons, shared by many others. The atmosphere is the ultimate commons: everyone uses it, and everyone is affected by it.

The problem is, none of us own it, so we tend to slide into the natural tendency to do what's best for us in the short term. There are ways to reduce carbon dioxide, to slow down the influences of climate change until other solutions can be developed. None of these choices are easy, and many require a significant lifestyle change.

## Reducing Carbon Dioxide Emissions

As the old expression goes, "An ounce of prevention is worth a pound of cure." In the long run, the cheapest and most effective way to deal with global climate change (and most other environmental issues) is to prevent it from happening.

There are a lot of aspects of our atmosphere and climate system that are out of our control. Emissions of carbon dioxide and methane, on the other hand, are not.

You may have heard of the term carbon footprint. This is a measurement of how much carbon dioxide an individual or company produces over the course of a year. How do we reduce this? Remember, the greatest source of $CO_2$ emissions is the burning of fossil fuels. Reducing your carbon footprint means using less electricity (burns coal), lessening the amount of driving you do (burns gasoline), and turning down the furnace in your house during the winter (burns natural gas). Conservation efforts like this are going to be an important part of the solution, at least until alternative energy sources like wind and solar become more widespread and affordable.

What about methane? Methane actually traps and releases heat more effectively than carbon dioxide does. Methane is a natural gas, and although we don't intentionally release it, it can certainly leak out of natural gas wells and pipelines. Reducing our natural gas consumption will help to lessen this source. Livestock also produce a lot of methane, a byproduct of their digestive systems. Millions of tons are released annually in the process of raising animals. Some movements, like Meatless Monday, seek to reduce the amount of meat we consume. This would reduce the overall size of global livestock herds, decreasing methane production.

## Regulating Carbon Dioxide

Changing lifestyle choices for individuals, especially those in developed countries, is an uphill battle. Perhaps, then, the better strategy would be to control greenhouse gas emissions from the biggest point sources.

One proposed idea is the carbon cap-and-trade system. This would work by instituting limits on how much carbon dioxide any given factory is legally allowed to produce. Carbon dioxide would be regulated, much like other air pollutants defined by the Clean Air Act. Some facilities, such as coal-burning power plants, would inevitably exceed these legal limits. They would have to purchase additional credits or permits to remain legal. Depending on how the system is designed, the credits may be bought from other businesses that don't need them or purchased auction-style from the government.

However the system would be designed, it would institute a financial incentive to businesses and industries to reduce their carbon footprint. The side effect of this is that some of these costs would be passed along to the consumer. Electricity from coal would become more expensive. Is that a bad thing? Perhaps, but remember the idea of *hidden cost* from Chapter 10. Pricing electricity in this way would more accurately reflect the long-term environmental costs of burning coal.

## Sequestering Carbon Dioxide

Reducing carbon footprints and switching to cleaner sources of energy would help to prevent additional carbon dioxide from entering the atmosphere, but it would do nothing about what's already there. This is where carbon sequestration comes into the picture.

Sequestration is any method of removing and locking carbon dioxide out of the atmosphere. Trees and other plants already do this through photosynthesis. A company called Ecometrica performed some measurements and calculations on a mature sycamore tree, finding that it was about 50 percent carbon by mass, or 1 ton. If this tree were burned, the carbon would combine with oxygen to form more than 3 tons of carbon dioxide. All that, in just one tree!

Research is also being conducted to see if carbon dioxide can be captured and injected deep underground. A federally funded demonstration project in Illinois captures $CO_2$ from an ethanol plant, compresses it, and then injects it deep into a layer of sandstone. The long-term success of these projects is yet unknown.

## Geoengineering

Depending on how quickly and drastically the Earth's climate begins to change, we may have to resort to some pretty drastic measures. Inspiration for one of these measures comes from volcanoes—a factor behind many of the Earth's climate changes in the past.

**CASE STUDY**

An episode of the television show *Futurama* featured a mock global warming public service announcement. This tongue-in-cheek prediction was made about how the problem was handled: "Ever since 2063, we simply drop a giant ice cube into the ocean, every now and then. Of course, since the greenhouse gases keep building up, it requires more and more ice each time. Thus, solving the problem once and for all."

This is an example of geoengineering—intentionally altering the Earth's climate system specifically to moderate the effects of global warming. The idea is to inject particles of sulfur-containing gases into the stratosphere. These particles would partially block some of the incoming solar energy and hopefully offset the increased greenhouse effect.

What could possibly go wrong? Well, remember, the climate system is very complex. Countless variables affect it. There's a reason why weather forecasts become less and less accurate three or more days out and nearly completely unreliable more than 10 days out. Cloud formation could be affected. The ozone layer could be impacted. A climate simulation run by environmental scientists at Rutgers University showed that the summer monsoons that southeast Asia depends on could be disrupted.

The layer of greenhouse gases in the troposphere traps and returns heat emitted from the Earth's surface. Human activity has increased the amount of carbon dioxide within this layer. Heat in the atmosphere and oceans drives the water cycle, which is a big part of our climate system. Put all these pieces in place, and anthropogenic climate change is a huge, looming issue for our society.

A combination of strategies will be necessary to manage this problem. Preventative strategies, such as energy conservation and cap-and-trade systems, will lessen the effect of global warming, while cleanup strategies seek to remove what carbon dioxide is already present. None of these changes will be easy or cheap. They will all require a significant investment and a major lifestyle change, especially on the part of the developed world.

CASE STUDY

How much of a debate really surrounds global warming? If you tune out the mass media and only look at the scientists, the answer is a lot less than you might think. A professor at the University of Illinois surveyed a group of 3,146 Earth scientists about their opinions on whether the Earth is warming and whether humans are behind it. The answer varied, depending on the group. Climatologists were nearly unanimous, with 97 percent agreeing with the two statements. Only 47 percent of petroleum geologists agreed. Overall, 82 percent of the scientists agreed with both statements.

## The Least You Need to Know

- Peer-reviewed data supports the conclusion that increases in global surface temperatures are primarily due to anthropogenic greenhouse gas emissions.
- Greenhouse gases contribute to the ability of the atmosphere to retain and emit infrared (heat) energy that escapes from the surface.
- The effects of climate change are widespread, including the altering of ecosystems and amplification of major drought and flood events.
- Global climate change is often painted as a debate or uncertainty, as efforts to deal with it will be costly and require big lifestyle changes.
- Preventative strategies, such as emissions restrictions, keep carbon dioxide from being released. Cleanup strategies, like sequestration, remove what is already present in the atmosphere.

# Waste

As long as humans have walked the Earth, we've produced waste. In the beginning, during the time of the hunters and gatherers, the waste was no different than that of any other animal. Food scraps, feces, and the bodies of the dead would be left behind as tribes moved about. These would all be recycled back into the soil or atmosphere through the action of bacteria and fungi.

Following the Agricultural Revolution, people began living in permanent villages. The amount and types of waste began to change. Today, when archaeologists search for the remains of these past settlements, they look for elevated areas called "mounds."

According to a study by Charles Gunnerson, many ancient cities existed at a somewhat higher elevation than their sur-rounding geography—anywhere from 10 to 400 centimeters. This was due to the ancient practice of discarding waste on the floors of their homes, then burying it with a fresh layer of dirt and clay. While this practice no longer exists, the amount of waste we produce is far greater and comes with a lot more complications.

## In This Chapter

- The different types of waste produced by human society
- How sanitary landfills keep waste securely locked away
- The advantages and disadvantages of incineration
- Methods of reusing, recycling, and reducing waste

# The Human Waste Stream

Dealing with waste is a complex puzzle. It's a mixture of many different types of materials, coming from many different sources. Each has its own set of properties, rate of decomposition, and health risks.

The sum total of all this waste is called the *waste stream*. Imagine it this way: take the entire human population and concentrate it into one giant megalopolis. Now visualize all the waste produced by these 7 billion people. Food scraps, paper, plastics, glass, everything, all flowing out in a single massive river ... of garbage. If you were charged with managing this waste stream, what would you do? Bury it? Burn it? Sort it and try to reuse it somehow? This gives you an idea of the scope of the problem.

## Agricultural and Mining Waste

Not all waste our society produces is in trash cans. Some of it is a natural byproduct of growing food or extracting fuels and mineral resources.

Agricultural waste is all of the leftovers after a harvest has been completed. This could range from inedible plant matter, such as corn stalks, to animal manure and unused fertilizer. Fortunately, most of this material is *biodegradable* and will naturally decompose over time, as opposed to *nondegradable* waste.

**DEFINITION**

The **waste stream** is the total production of waste from human society, including agricultural, industrial, municipal, and mining sources. **Biodegradable** waste can be naturally broken down by decomposers in the environment, while **nondegradable** waste persists for centuries.

Mining waste is a little different. This is the material that has been brought up from the Earth in the process of extracting coal, metal ore, or some other resource underground. Mining waste technically consists of materials naturally found in the Earth's crust, but they're now exposed to the wind and rain. The biggest problem with mining waste is usually erosion and runoff.

## Municipal Solid Waste

Municipal solid waste is what we commonly refer to as garbage. Garbage is a challenging form of waste to deal with because its composition is so varied; there are so many different things found inside it.

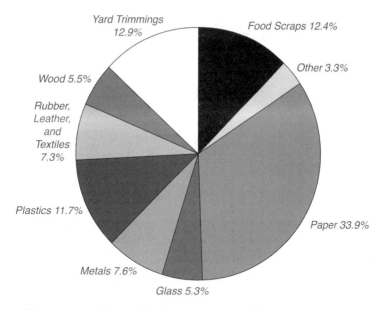

*The components of the municipal waste stream, according to a 2006 EPA report.*

Municipal solid waste is made of materials that are inert—they don't present an immediate danger of chemical reaction or explosion. Some of the materials are biodegradable, like yard waste and food scraps. Others don't degrade, like plastics, glass, and metal. They essentially remain in the form we cast them in for thousands of years. When a garbage truck leaves a neighborhood, it contains all of these sources and more.

 **CASE STUDY**

> Old landfills that weren't lined properly are a significant source of groundwater contamination. Two of the most dangerous toxins produced are vinyl chloride and dioxin. Both are considered carcinogenic, and have been linked to health problems dealing with hormones and the reproductive and immune systems.
>
> Millions of dollars have been spent excavating these old landfills and retrofitting them with modern liners and caps to safely seal them, at least for the time being.

Municipal solid waste is also one of the fastest-growing parts of the waste stream. According to the EPA, production has increased from 88 million tons annually in 1960 to 250 million tons currently.

## Hazardous Waste

The last component of the waste stream is the most acutely dangerous. Hazardous waste is defined as any discarded liquid or solid that meets at least one of the following criteria:

- Fatal to humans or lab animals in low doses.

- Known to be a carcinogen or teratogen.

- Will catch fire and burn at temperatures less than 60°C.

- Corrosive to metals or plastics.

- Reacts violently with other compounds found in waste.

According to the EPA, the United States produces over 200 million tons of hazardous waste per year. Not all of it comes from some nameless factory, either; there are some household wastes that are hazardous, such as batteries, paint solvents, and used motor oil.

# Nonhazardous Waste Disposal

Now that we have an overview of exactly what is in the waste stream, now comes the hard part: what to do about it. Again, we're talking about a pretty diverse mixture of substances here, so let's at least separate it into hazardous and nonhazardous waste.

Nonhazardous waste is mostly municipal solid waste. You know, garbage. Remember the ancient cities that would simply dump their trash onto the floor of their homes and then bury it with a layer of dirt, causing the city to gradually rise over decades? Well, we haven't changed as much as you might think.

## Open Dumps

Open dumps are the simplest, cheapest, and fastest way to deal with garbage. A community will have a shared, open area where everyone can leave their garbage. These areas usually are far enough away from the town that nobody has to actually look at it (or smell it).

The downside to this kind of waste disposal is obvious. Open dumps leave the waste completely exposed to the elements. If it rains heavily or floods, some of the waste is likely to infiltrate and contaminate the nearby water supply.

Pest animals are attracted to the smell of decomposing garbage. Everything from flies to seagulls and bears will descend upon the dump. As the community grows, so does the dump. Eventually a new area may have to be cleared, and the problem spreads. Clearly, this is a completely unsustainable approach.

# Bury It

The next option is to take the waste and simply bury it in a *sanitary landfill*. This keeps it out of sight, limits the number of scavenger animals, and reduces the smell. There's one big problem with landfills, however. When it rains, the water will percolate through and come out the bottom as *leachate*.

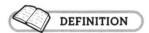

**DEFINITION**

> A **sanitary landfill** is a large area that has been excavated and lined to allow the safe deposit of municipal solid waste. These landfills are monitored closely for the presence of **leachate,** contaminated water that percolates through the landfill and exits the bottom.

Leachate is a nasty, toxic mix of chemicals picked up from all the different materials throughout the landfill. It could contain partially decomposed organic matter, bacteria, and chemicals released from degraded synthetic material like plastic. This isn't anything you want to get into your groundwater, but that's exactly where it heads.

Modern sanitary landfills employ protective measures to prevent this contamination. First, a clay and plastic lining is placed along the bottom to catch the leachate. A system of pipes will be used to drain the leachate out as it collects. It can then be treated before it's discharged.

Other than the leachate, sanitary landfills are a fairly simple operation. Municipal solid waste is trucked in daily, where it's dumped and compacted by large vehicles. A layer of dirt is placed over the garbage daily, to reduce odor and give the machines a solid ground to work on.

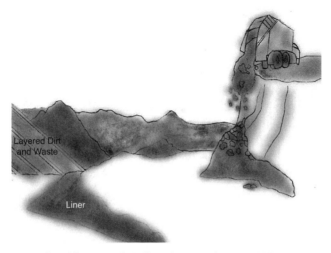

Layered Dirt and Waste

Liner

*Sanitary landfills see a daily inflow of municipal waste, which is compacted and covered by a layer of topsoil.*

As long as the leachate is contained, the landfill seems like a good solution. However, it's only a temporary solution. Eventually, the landfill will run out of space. Finding a suitable space to build the next one, especially as urban areas spread, is extremely difficult.

## Burn It

As landfill space becomes scarce, many communities have had to find an alternative way to reduce their waste volume. The answer has been *incineration.*

Burning waste is very effective at reducing its volume, anywhere on the order of 75 to 90 percent. This eases pressure on landfills, since less volume means less space is needed. Incinerators do have a few drawbacks. First, they're expensive to build and operate. As you'll remember, according to the Law of Conservation of Mass, matter cannot be created or destroyed. The lost volume of the waste ends up in the air. A series of air pollution controls must be installed to keep pollutants like *fly ash* out of the air.

The other problem is the leftover *bottom ash.* While the overall volume of the waste has been effectively reduced, the ash is more concentrated and toxic. Remember, it contains everything that wasn't burnable in the incinerator. Some space in a sanitary landfill will be required to safely contain this ash.

 **DEFINITION**

**Incineration** is the process of burning municipal solid waste to reduce its volume. The burnable components of the waste become gas, while the rest is reduced to ash. **Fly ash** is the component of that ash that actually becomes airborne in the process of incineration. **Bottom ash** is the solid portion that remains and must be buried in a landfill.

# Hazardous Waste Disposal

Not all wastes can be handled with landfills and incineration. Some types, particularly industrial wastes, are so highly toxic or reactive that special attention must be given to their disposal.

Like many of the environmental lessons we've learned as a society, this one had to come the hard way. Before the modern environmentalism movement, before the EPA was established in the 1970s, there were very few laws dealing with the disposal of waste of any kind, let alone hazardous waste.

## A Bad Burial

The most egregious example of how not to deal with hazardous waste begins with a man named William Love, who had a vision of a city that was powered entirely by hydroelectricity. In the 1890s, he began to dig a canal that would connect to the Niagara River near the falls. The canal would carry water to a site where a dam could be constructed and his town would be built.

But the project was abandoned as the United States entered an economic depression and advances in electricity allowed it to be transported over greater distances. The end result was an empty mile-long, 50-foot-wide canal.

The land was eventually bought by Hooker Chemical, a company that produced a wide range of chemical products. They used the canal as a convenient dumping site for a wide variety of industrial wastes, many of which were hazardous. Once full, the canal was sealed with a clay and dirt cover.

As the nearby city of Niagara Falls grew, pressure was placed on Hooker Chemical to sell the land for residential development. Hooker didn't want to make the sale but eventually consented in 1953 under the condition that the city agreed it understood what the land had previously been used for.

Homes were constructed around the canal, and an elementary school was built right next to it. Disaster struck the area when the meltwater from a heavy winter seeped into the canal, causing some of the liquid waste inside to ooze up into basements, yards, and playgrounds. Local children became afflicted with mysterious ailments. Eventually, the town was evacuated.

*A map of the Love Canal development and elementary school immediately adjacent to the dump site.*

## Containment and Remediation

Soon after the Love Canal disaster hit the headlines, two important laws dealing with hazardous waste were passed. The first, the Resource Conservation and Recovery Act (RCRA) of 1976, required any company producing hazardous waste to treat and dispose of it safely. This included a "cradle-to-grave" accounting of what happened to the waste step by step, until it was safely secured in a landfill.

The second law was the Comprehensive Environmental Response, Compensation, and Liability Act (CERCLA) of 1980, more commonly known as the "Superfund Law."

The Superfund Law gave the EPA the ability to identify any party responsible for the creation of a hazardous waste site and force them to clean up the site or help fund its cleanup. This is what eventually happened with the Love Canal site. A new liner was placed over the top of the entire canal, and a collection system drains the leachate and treats it before returning it to the local water system. People have even begun living in the area again.

## Someone Else's Problem

Another method of dealing with hazardous waste was to ship it elsewhere. Developed countries would export their waste to poorer countries, paying them a fee to accept it.

During the 1980s, the city of Philadelphia experienced a severe shortage in its landfill space. As a result, thousands of tons of incinerator ash built up without a place to go. A private contractor agreed to load the ash onto a barge called the *Khian Sea* and find a suitable dump for it.

The *Khian Sea* ended up going on a two-year voyage, visiting multiple countries in Central America, the Middle East, and Southeast Asia before the crew eventually just dumped the ash into the Indian Ocean. Many developed countries (but not the United States) later signed and ratified an international treaty called the Basel Convention, agreeing to end the practice of exporting waste to other countries. While the United States signed on to the treaty, it was never ratified over concerns that it could interfere with the export of scrap materials like steel.

 **CASE STUDY**

Electronic waste is one of the fastest-growing components of the modern waste stream. Obsolete electronics, like cell phones and computers, contain large amounts of toxic material such as heavy metals. These components are often smuggled overseas, where destitute workers will disassemble them by hand so the metals can be sold and reused. The work is dangerous and toxic, causing a series of health problems among the workers and greatly contaminating the local soil and water supplies.

# Slowing the Waste Stream

All of the methods of waste management we've covered so far have their share of problems. Landfills are temporary solutions. They fill up and must be carefully monitored for the remainder of their existence. Incinerators reduce waste volume but are expensive and polluting.

The best way to deal with waste may be to mimic nature. Remember from Chapter 8 that matter in natural ecosystems cycles. Carbon, sulfur, water, and the various other components of organic matter are constantly recycled and reused by other organisms.

In human waste management, there's a term for this idea. It's called "cradle-to-cradle" design. The idea is that any material that's produced, no matter what the purpose, can be collected and used again and again. In a perfect world, this kind of design would completely eliminate the need for landfills and incinerators.

## Reducing

With many of the environmental problems we've discussed, the cheapest and easiest solution is to not produce it in the first place. This is the case with water pollution, air pollution, and even with global climate change. In the long run, prevention is always less expensive than cleanup.

With this idea in mind, think about what the biggest component of your weekly trash pickup is. If your garbage can is nearby, take a look inside. What makes up the majority of the material? The answer is probably packaging. Cans, bottles, boxes, or plastic trays make up a huge proportion of municipal solid waste. This is why solid waste production has tripled since the 1960s!

One of the best ways to reduce waste is to adopt lifestyle changes that use less packaging. Instead of drinking bottled water, install a filter in your sink and carry around a reusable bottle. Rely less on pre-packaged foods and cook more raw meat and produce.

## Recycling

A few lifestyle changes can greatly reduce one's production of garbage, but it won't eliminate it entirely. That isn't necessarily a realistic goal for everyone, and some waste production is inevitable.

This is far from a disaster. How much of the trash we produce can be recycled into new products? Many materials can, but the easiest ones to start with are glass, metal, paper, and plastic.

Depending on the community, just about any kind of glass can be collected and melted down into new products. The same can be said of steel and aluminum cans. Paper, cardboard, and paperboard also can be recycled, as long as they aren't contaminated by oils or grease (sorry, no pizza boxes).

Recycling is a great way to prevent materials from entering landfills and incinerators. It also greatly reduces the environmental impact of many products. Recycling aluminum, for example, saves the cost and damage of mining and processing raw ore from the ground.

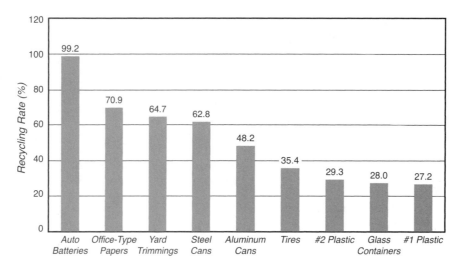

*As this 2008 data from the EPA shows, recycling rates for the different components of the municipal waste stream vary widely.*

One of the biggest problems with recycling is that not all the materials are valuable enough to provide an incentive for their collection. Aluminum and steel are pretty valuable, but plastic is so cheap that many communities simply cannot financially support its recycling. Check with your local municipality to see what products are accepted.

 **A LOOK BACK**

The first state in the United States to impose a container deposit was Oregon in 1972. A five-cent deposit added to every beverage container could be recollected when the container was returned to a recycling facility. Many other states have adopted similar deposits, leading to a huge increase in container recycling rates. Beverage companies oppose this legislation, concerned that the added cost may dissuade consumers from purchasing their products.

## Composting

Recycling works for some materials, but what about the biodegradable organics? These can be recycled, too, but in a different way.

Fallen leaves, grass clippings, and food scraps are all organic—they derive directly from living organisms. As such, they can be broken down naturally by decomposers like fungi and bacteria.

Large, commercial composting operations have massive grinders that mix up food scraps and put them in huge black bags that absorb sunlight and maximize the rate of decomposition. But even small households can have their own compost pile in the backyard.

## Bioremediation

What about hazardous waste? Is there any way to deal with this material in a manner more effective than simply locking it away in a landfill? One possible answer is bioremediation—harnessing living organisms to absorb and metabolize toxins.

There are many different species of plants and fungi that will naturally absorb and store specific toxic compounds found in the soil or water. Willow trees, for example, are known to absorb the metals cadmium, zinc, and copper. Sunflowers have been used to absorb radioactive isotopes in the Chernobyl exclusion zone!

A recent discovery was made at Yale University of a species of fungus that can digest polyurethane—a component of many plastic products. Imagine if we could decompose the massive accumulations of plastic in the world's landfills with a humble fungus!

## The Least You Need to Know

- The two biggest parts of the waste stream that we deal with are municipal solid waste (garbage) and industrial hazardous waste.
- Sanitary landfills are effective at locking away wastes and separating them from communities, but they do fill up over time.
- Incinerators greatly reduce the volume of trash, but are very expensive to build and maintain compared to landfills.
- The cheapest and easiest method of dealing with waste is to not produce it in the first place. Other techniques like recycling keep it from entering the landfill or incinerator.
- Bioremediation is a technique that employs living organisms like plants and fungi to absorb and metabolize toxins.

# Hope for the Future

In Part 6, we take an optimistic look at specific changes and technologies that are vital to the human population finding a way to live on Earth sustainably.

The first step is managing and reducing the massive amount of energy we consume. Technologies to encourage conservation, like the smart grid, are coupled with innovative technologies that improve efficiency, like hybrid cars. Dealing with our energy use will go a long way toward dealing with all the issues of pollution and global climate change.

The next step is to explore ways to live in the environment while preserving it as much as possible. Urban sprawl is causing a slow but steady loss in open land, requiring us to invest more in national parks and wildlife refuges. The ultimate goal of environmental science is sustainability.

I conclude by offering some thoughts on changes in our society—from the design of our cities to our own lifestyles—that will be necessary to achieve sustainability.

# Reducing Energy Consumption

We've covered a lot of ground in this book, including the damage to the environment that has occurred as a result of human activity. A pessimist might look at our society and describe it as unsustainable and ultimately doomed.

This viewpoint does not consider the tremendous power of human ingenuity. Look back at our history and the number of hurdles we've overcome in growing to a population of 7 billion. Modern water treatment has reduced outbreaks of intestinal disease to a rarity, at least in the developed world. Vaccines and antibiotics have staved off premature deaths from viral and bacterial disease.

As our population continues to grow along with our demand for natural resources, something is going to have to change so we can survive in the long term. That change will have to occur in both our technologies and our lifestyles.

## In This Chapter

- How the "smart grid" changes the way electricity is metered
- The importance of energy storage technologies
- How energy efficiency is determined
- The relative efficiencies of hybrid and electric vehicles

# The True Cost of Energy

All of the environmental problems we've covered to this point are complex and have many different causes. However, there is one common source, one societal need that impacts the environment to a greater degree than any other: energy.

Think about all the different ways we use energy: for transportation, electricity, and heat. Now, recall all the different sources we have for getting that energy: fossil fuels, renewable sources, or nuclear power.

Some of the most significant instances of air and water pollution that have occurred were directly tied to energy. Oil spills in the ocean, the deadly smog of London and Donora, the irradiation of hundreds of square miles surrounding Chernobyl. If we can better manage our extraction and use of this resource, each of these problems can be significantly reduced.

## The Smart Grid

As we covered in Chapter 10, money is an important driving force behind many of the decisions made by human society. If the way we use energy is to change, there probably will have to be a financial incentive to do so. One possible way is to change our electricity infrastructure.

This is the smart grid, a series of innovations to improve the efficiency and reliability of our electricity delivery system. Specific features of the smart grid would include:

- The ability to handle bidirectional energy flows: to and from homes.
- Automatically adjusting household thermostats.
- Charging consumers different rates for peak and off-peak energy consumption.
- Adjusting electricity production in response to the variability of solar and wind farms.

Practically, what would these changes mean for you? If you installed solar panels on your roof and generated more electricity than you use, the power company could collect the electricity and refund the difference. During periods of high load, such as hot summer days, household thermostats would be automatically adjusted upward so overall energy consumption was lower. This would prevent brownouts and blackouts. Charging different rates for electricity would encourage consumers to cool their homes at night, when the overall load is less.

## Microgeneration

One of the intriguing ideas of the smart grid is the ability of households to generate their own electricity and actually be paid for it. This concept is known as *microgeneration*.

Producing your own electricity has quite a few benefits. The first, and most obvious, is that you would save money over the long run. Of course, installing the system would be a significant initial investment. Over time, however, the costs could certainly be recovered.

Second, the electricity doesn't have to be transmitted long distances. This means less stray electricity is lost by the power lines, and the consumer is less affected by blackouts.

## Energy Storage

The last component of the smart grid system is energy storage. One of the disadvantages of utilizing renewable sources like wind and solar is that they are variable—the wind doesn't consistently blow at the same speed, and the sun isn't always shining.

 **CASE STUDY**

The Energy Storage Challenge is a competition that started in 2011 for startup companies to propose innovative ideas for energy storage systems. The winner of the first competition that year was the United Kingdom–based Cella Energy, who proposed a hydrogen-fuel storage system. The fuel would have a greater storage capacity than even the best batteries available and produce no emissions when combusted, with the exception of water vapor.

This problem could be solved if surplus energy could be captured and stored until needed. We already do this on a smaller scale with batteries. The problem is trying to apply this idea to entire households or even cities. Consider the battery technologies that currently exist:

- Alkaline batteries are not rechargeable.

- Lead-acid (car) batteries are too big and heavy.

- Nickel-metal hydride (NiMH) batteries have a tendency to lose their charge over time.

- Lithium-ion batteries are very expensive to manufacture.

There are some alternatives. Some hydroelectric dams have pumped storage generating systems. These are separate reservoirs that water is pumped into during off-peak hours. The water then can naturally (through gravity) pass back through the dam, and the energy is recollected by the turbines.

# Improving Energy Efficiency

Part of the answer in dealing with our energy problems is to develop better, cleaner sources. Another piece to the puzzle is to try to use less energy. Naturally, people could simply try to adjust their lifestyles and make do with less. However, it's difficult to make this adjustment, especially when a society becomes accustomed to a certain lifestyle. With this in mind, improving energy efficiency might be the solution.

Energy efficiency is actually a ratio of the amount of energy put into something (i.e., a machine or appliance) and the amount that is "useful." This means the energy is actually applied toward its intended purpose.

For example, let's say you drive your car on a quick errand. The car uses about 300 kilojoules of energy. However, only 45 kilojoules actually go into moving the car forward. Efficiency can be calculated like this:

$$\text{Efficiency} = \text{Energy out} \div \text{Energy in} \times 100 \text{ percent}$$

$$\text{Efficiency} = 45\text{kJ} \div 300\text{kJ} \times 100 \text{ percent}$$

The car is only 15 percent efficient! This sounds pretty lousy, but remember from Chapter 8 that even ecosystems follow the 10 percent rule. Where does the rest of the energy go? Most of it is given off as waste heat. This is why the engine gets so warm. If we can improve the efficiency of the different machines and appliances we use, we can reduce our total energy requirement.

## The Green Home

The first step in improving efficiency begins at home. One of the biggest uses of energy in the home is heating and air conditioning. Heating costs can be lowered by improving insulation, installing windows that allow less heat to escape, and installing more efficient furnaces. According to the EPA Energy Star website, the average home can save 10 percent of its annual energy bill just by sealing and insulating better.

As technologies improve, furnaces and air conditioners have become increasingly efficient. For example, newer furnaces are 95 percent efficient, compared to the 80 percent models that have been standard in the past.

Finally, installing a programmable thermostat can help to control heating and cooling costs. These thermostats allow you to set what the household temperature will be at different times of the day. If you're at work during the day, for example, you could set the thermostat to a much lower temperature. It would then automatically reset back to the normal level when you got home.

## Hybrid Vehicles

Controlling household heating helps to reduce our need for natural gas. Decreasing air conditioning in the summer takes pressure off coal-fired electricity power plants. To round out our reduction in fossil fuel consumption, we need to deal with crude oil, and our biggest use of crude oil is for gasoline.

For many people, driving is a necessity. Public transportation might not be available, or it simply doesn't fit their work schedule. How, then, can we reduce our demand for gasoline? Again, the answer might be to improve efficiency.

For decades, cars have run on the same basic type of engine: internal combustion. Huge improvements have been made since the oil crisis of the 1970s to get the most out of these engines.

One big advance that has emerged in the last few years is hybrid cars. Hybrid cars have two energy sources: a traditional internal combustion engine and a large, rechargeable nickel-metal hydride or lithium-ion battery. The cars switch to battery mode while idling or coasting. They're also able to recapture some energy normally lost, such as during braking. As a result, these cars demonstrate remarkably higher gas mileage.

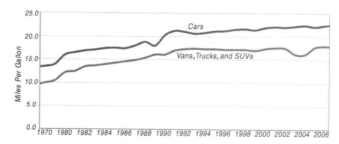

*According to the U.S. Energy Information Administration, gas mileage steadily climbed since the energy crisis of the 1970s, although it plateaued during the SUV craze of the early 2000s.*

## Electric Vehicles

Hybrid cars are a good step toward reducing energy use, but they ultimately are still reliant on a fossil fuel, and all of the bad things that come with extracting and burning that fuel.

The next step is likely to be the fully electric car. Electric cars are even more efficient than hybrid vehicles. They also have the added advantage of not being reliant on any one source of energy. Since they are electric, they can be recharged from just about any energy source, from coal to hydro or wind power.

One of the biggest drawbacks to electric cars right now (besides the cost) is that the infrastructure of most countries is not designed for this type of vehicle. We have gas stations, not recharging stations. There's going to have to be a gradual turnover of how we deliver energy to our vehicles as the technology changes.

As a consumer, trying to keep track of emerging technology and energy efficiency can be a daunting task. One program designed by the United States EPA in 1992 actually attempts to make this a little easier: the Energy Star program. For an appliance to qualify for the Energy Star mark, it must be more efficient than what is federally required. The actual amount of improvement varies by appliance, but is around 20 to 30 percent.

*The Energy Star service mark can only be placed on products that exceed the energy efficiency of their average counterparts.*

## Light Bulbs: An Efficiency Story

A good example to summarize the power of making a few improvements to efficiency is the humble light bulb. Light bulbs have gotten a lot of attention because everyone uses them. In fact, between ceiling lights and lamps, a typical household may use dozens of light bulbs.

Incandescent light bulbs have been around the longest. They pass electric current through a metal filament, which glows and produces light. As anyone who has ever touched a light bulb knows, they also produce heat. A lot of heat. In fact, traditional incandescent light bulbs are only about 10 percent efficient. They waste a lot of energy.

In 2007, the U.S. Congress passed the Energy Independence and Security Act. It contained a set of efficiency standards for all light bulbs, including incandescent. The sale of all 100W and 75W light bulbs would be banned as of 2013. This law caused a great deal of controversy within Congress as the date approached, and funding was pulled from the Energy Department to prevent the regulations from being enforced. Opponents of the law felt that the cost requirements demanded of consumers and businesses were too great and were not justified by the energy savings.

There are now some additional choices in light bulbs. Before we compare them, however, you have to understand that different light bulbs have different levels of brightness, measured in a unit called the *lumen*. A 60-watt light bulb, for example, produces 800 lumens of light. To make a valid comparison, we need to compare light bulbs that produce the same brightness.

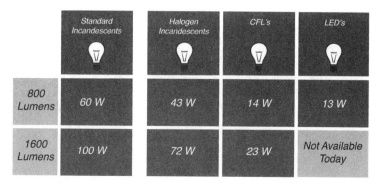

| | Standard Incandescents | Halogen Incandescents | CFL's | LED's |
|---|---|---|---|---|
| 800 Lumens | 60 W | 43 W | 14 W | 13 W |
| 1600 Lumens | 100 W | 72 W | 23 W | Not Available Today |

How much of a difference does switching a light bulb really make? An LED light bulb that produces 800 lumens of light only uses 13 watts of power, compared to 60 watts in a standard incandescent bulb. Let's assume that the light bulb is in use 3 hours a day, and the price of electricity is 13 cents per kilowatt-hour.

A standard incandescent light bulb?

> 60 watts × 3 hours × $0.13 per kilowatt-hour = $23.40 per year

A new LED light bulb?

> 13 watts × 3 hours × $0.13 per kilowatt-hour = $5.07 per year

Huge difference, right? The savings are substantial, not just financially, but in energy consumption. Now, imagine that every light bulb in use across the entire country was switched. The problem is, LED light bulbs are expensive. They might cost upwards of $20 each, while a traditional incandescent is under $1. Sure, you recover that cost over time, but most people don't have $20 to drop on a single light bulb!

Reducing energy consumption will not be a simple task. Over time, improved technology will spur the replacement of older appliances with more efficient ones. This will be a slow process, though, because investing in these new technologies is expensive. If we are able to transition fast enough, a combination of improved efficiency and a switch to less-polluting energy sources may slow some of the damage of pollution and climate change.

## The Least You Need to Know

- Improvements to current energy infrastructure are necessary to accept household-generated electricity.
- Smart electric meters will have the ability to charge different rates for electricity during peak and off-peak times, and automatically adjust thermostats.
- One of the limiting factors in using renewable energy is the lack of energy storage technology.
- Machines with greater efficiency take more of the input energy and apply it toward its intended purpose.
- While hybrid cars reduce the need for gasoline, they don't eliminate it. Totally electric cars can be charged from any energy source.

# CHAPTER 24

# Sustainable Land Use

In Chapter 1 of this book, we covered the three major environmental problems: resource depletion, pollution, and loss of biodiversity. In the last chapter, we tried to find ways to mitigate the damage caused by our need for energy. If we were successful, this would have a significant impact on both resource depletion (fossil fuels) and pollution. What about biodiversity?

The number and frequency of extinctions have increased alarmingly in the last century. The reasons behind extinctions vary from case to case, but many of them can be traced back to the expansion of our population.

Why should we be concerned about biodiversity? Some species of animals and plants are useful to humans. They help to provide food, control pests, or even contain chemical compounds that can be used to treat disease. Additionally, though, isn't there an ethical problem here as well? As the most prolific mammals on the planet, do we have a responsibility to allow other species to exist in their own right?

## In This Chapter

- The human-related causes of extinction
- The importance of protecting ecosystems and biodiversity
- How open land is preserved and protected
- Laws that protect endangered species

# The Next Great Extinction

*Extinction* is a natural phenomenon and can occur under many different conditions. An ecosystem can suddenly change, or the climate can shift, removing a niche that a species once occupied. The enormous insects found during the Carboniferous period no longer exist, because the atmosphere doesn't contain enough oxygen to support them.

Ecosystems are not now, nor have they ever been, permanent. Continents drift, and the climate shifts. This is a normal part of the Earth's history. What's happening now, however, is a series of extinctions that largely revolve around the growth of the human population.

## Ecosystem Fragmentation

The single biggest cause of *anthropogenic extinction* is *habitat degradation*. A whole host of human influences can degrade a habitat. The introduction of toxic pollution, for example, can render an area of land or water uninhabitable for certain species.

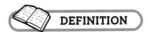 **DEFINITION**

> **Extinction** is the complete loss of a species from the Earth. **Anthropogenic extinction** is directly the result of human actions. The biggest cause is **habitat degradation,** which occurs when a habitat is changed in such a way that it no longer supports the species.

One of the greatest problems stems from a phenomenon called *urban sprawl.* A large population of people has migrated out of the city and into outlying areas, especially as transportation has improved. From their perspective, this grants them a place to live that's quieter and has more open land. From the perspective of an ecosystem, however, this is an extraordinarily destructive practice.

Consider the entire infrastructure that goes into a suburban area: highways, railroads, water lines, parking lots, and buildings. Once built, each of these structures occupies an area that was once a habitat for many different species. In addition to that, these structures cause *habitat fragmentation.*

A given ecosystem can be divided into two sections: an *interior habitat* and an *edge habitat.* The characteristics of the ecosystem are a little different in each area. In a forest, for example, the tree growth might be thicker in the interior, with a taller canopy. The edge habitat might have more open fields.

When one of these ecosystems is divided, such as when a road is built through it, two things happen. First, a large, continuous habitat is divided into two smaller ones. Migrating between the two areas becomes much more difficult (and dangerous) for the animals that live within. Next, the amount of interior habitat is sharply diminished. Imagine building a road right through a large forest; that thick, tall interior must be cut down to accommodate the highway! As a result, you're left with more edge habitat. The very nature of the ecosystem is completely changed.

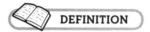

**DEFINITION**

**Habitat fragmentation** occurs whenever an ecosystem is divided in some way by a physical barrier, such as a road. This changes the dynamics of the ecosystem. The **interior habitat**, which has the densest growth of the most mature and long-lived species, is reduced in area. The **edge habitat** is found between two habitats or a habitat and a barrier. Pioneer species are more likely to be found here.

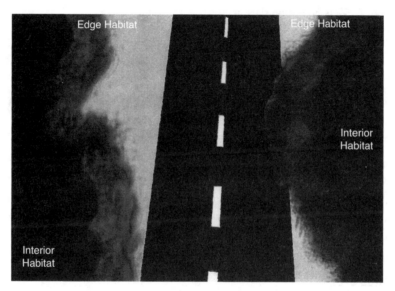

*Human structures such as roads can divide ecosystems, causing an increase in edge habitats and a decrease in interior habitats.*

Some organisms adapt well to change. These species tend to be physically smaller, reproduce quickly, and produce lots of young with each generation. They're called *r-selected species.* Insects and weeds would be good examples. Other organisms, called *K-selected species,* are much more vulnerable to this kind of rapid change. Mature trees and large mammals like the giant panda are good examples. These organisms are the ones most likely to face extinction as the result of habitat loss.

**DEFINITION**

The r/K selection theory divides organisms into two categories, based on certain characteristics. The smaller, shorter-lived, and more quickly reproducing are called **r-selected species.** These organisms tend to excel in changing environments. Larger, longer-lived organisms that have only a few offspring at a time are called **K-selected species.** These require a stable ecosystem.

## Invasive Species

In Chapter 1, we covered the ecosystem destruction in Australia caused by the introduction of the European rabbit. This organism is an *invasive species*—one that's not normally found in the ecosystem and lacks normal population controls (predators).

There's a huge list of invasive species living all over the world, and each has caused irreversible changes to the ecosystems they were introduced to. One of the best and earliest examples of this is the *Rattus rattus,* or black rat.

The black rat actually originated in India and Southeast Asia, but gradually spread throughout Europe by the first century c.e. They frequently stow away on ships, resulting in their spread to just about everywhere in the world.

In island and coastal ecosystems, black rats have a voracious appetite for seabirds and their eggs. In forest areas, they ingest seeds and small invertebrates. None of these effects could have fairly been anticipated by the travelers of the time, but they've provided some important warnings to us today. Now, if you want to cross borders with living specimens, you will need special permission from the government.

**CASE STUDY**

The black rat, a species that has spread throughout the world and caused the devastation of multiple species of sea birds and other organisms, is itself becoming endangered in the United Kingdom. It's the victim of competition from the larger brown rat, another introduced species! Unfortunately, the brown rat is still an invasive species, and has many of the same appetites as the black rat.

## Overhunting

The most direct form of anthropogenic species extinction is overhunting. Simply put, this is the practice of hunting and killing an animal at such a rate that its population is unable to keep up.

Many species of animals have gone extinct or become endangered as a direct result of unregulated hunting. A good example of this is the island of Madagascar. As an island country,

Madagascar has a staggering amount of biodiversity in its rainforest ecosystems. Yet the diversity of large animals before humans arrived thousands of years ago was likely even greater, according to the fossil record.

Included in the list of now-extinct Madagascar species are several different types of lemurs. Anthropologists have discovered preserved bones of these lemurs with sharp cuts and chop marks—characteristic of animals that have been hunted and butchered for food.

The animals most at risk for this kind of extinction, again, are the K-selected species. Remember, they're big and have long life spans. They simply don't have the capacity to recover their population quickly. As a result of all these human actions—hunting and habitat fragmentation, the remaining lemurs and other animals of Madagascar are still very much at risk.

# Protecting What's Left

Loss of biodiversity might feel so out of control that it's too late to change. This is absolutely not the case. Remember from Chapter 2 that the loss of open land was recognized even during the nineteenth century, and was a priority of President Theodore Roosevelt.

The westward expansion experienced by the United States during the nineteenth century was only the beginning. Urban sprawl has resulted in contained cities stretching out farther and farther. In some cases, the sprawl of neighboring cities has merged.

*The megalopolis of Boston, New York City, Philadelphia, and Washington, D.C., is referred to as "Boswash" or "Bosnywash." This map is based on an image from* An Outline of American Geography, *a U.S. Department of State publication.*

## Alternative Urban Growth/Planning

Smart growth is an alternative to the pattern of urban sprawl that has dominated the last century. The idea is to make communities more sustainable—to provide opportunities and resources for those who live there, reducing commuting.

Many suburban areas today tend to have concentrated residential, commercial, and industrial areas. Mixing these areas to a greater extent, especially the residential and commercial, could reduce the amount of time everyone needs to spend on the road to get to work and take care of their errands.

Smart growth also places more of an emphasis on public transportation, bicycling, and walking. Automobiles, while convenient, are the least fuel-efficient and require the most construction in terms of roads and parking places. Each neighborhood would also have some open space and farmland preserved, to maintain a greater connection to the environment.

## National Forests and Grasslands

Controlling population growth and sprawl is part of the solution, but how do we manage the land that people aren't living on? The easy answer might seem to set it all aside as protected land, but we do need some land to produce food and timber.

The national forest system was created by the Land Revision Act of 1891. These are publicly owned forests that are managed by the United States Forestry Service. These are not refuges; rather, this is land that's meant to be used commercially. Logging, with restrictions, is allowed within national forests.

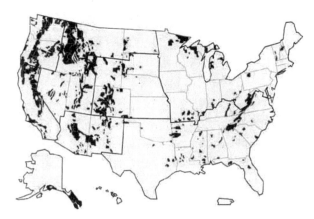

*The federally managed United States national forest system. Most of this land is found in the western states.*

There's plenty of room for debate over how these forests are managed. Environmentalists would prefer to see them preserved, but they mark an attempt to maintain the forests, and a movement away from the rampant clear-cutting that marked the first hundred years of the United States.

The Forest Service also manages the lesser-known national grasslands. These are handled in much the same way as the national forests—they can be used commercially, with permission, and under certain conditions. For example, some contain wind farms, while others are part of our rangelands: areas that cattle can be brought through to graze.

## National Parks

National parks are defined differently, depending on the country, but they all work on the same basic principle. They're areas of land that are conserved—areas where human use is very restricted.

### A LOOK BACK

When the land that would become Yosemite National Park was set aside and given to the state of California, the 38th U.S. Congress wrote, "... the said State shall accept this grant upon the express conditions that the premises shall be held for public use, resort, and recreation; shall be inalienable for all time." This planted the seed of what eventually grew into the United States National Parks.

In the United States, the first national park set aside was Yellowstone. The purpose was to allow recreational use, but not commercial. In other words, people could come in and boat, hike, and camp, but there wouldn't be any logging or hunting allowed.

## Wildlife Preserves

National parks and national forests are anthropocentric ideas—land managed for the long-term benefit of humans. Wildlife preserves (also called refuges or sanctuaries) are ecocentric. They exist to provide complete protection for the species that live within.

A good example of a wildlife preserve is the largest one—the Arctic National Wildlife Refuge (ANWR) in Alaska. Over 19 million acres of this land has been set aside, with no roads entering or leaving the refuge. The only people living within are scattered native settlements.

As with any ecocentric decision, there's often controversy with these preserves. The ANWR has been proposed as a site for drilling for crude oil on multiple occasions during the last several decades. The debate seems to arise whenever gas prices spike. However, the actual impact of the oil on the global markets would be small. According to the Energy Information Administration, maximum production on this land would only add up to about 1.0 percent of total world oil consumption.

## Wetland Remediation

Of all the ecosystems impacted by the growth and expansion of human civilization, wetlands may have borne the worst. Originally seen as wasteland, wetlands would be drained and used for land development purposes—turned into farms or towns. At one point, farmers were actually given subsidies to drain and use wetlands.

As you'll recall from Chapter 7, wetlands provide a lot of benefits. They help to reduce local flooding, increase aquifer regeneration, and are a major source of biodiversity on land.

The current policy for wetlands is best described as "no net loss." What this means is that developers can still drain and build on wetlands, pending government approval. However, they are required to build or restore an equivalent acreage of wetlands elsewhere to make up for this loss.

While this is a big improvement, wetlands are still very vulnerable. They're not included in the Clean Water Act, as they don't fit the law's definition of "relatively permanent water." Some have proposed this be changed, but as yet, they're not included within the law.

# Species Protection

Efforts have been made to begin protecting and preserving the remaining ecosystems on the Earth. What about the individual species within them?

Protecting every species from becoming extinct is a much more difficult task, as not all of the species on the Earth have been successfully identified yet. There may be countless extinctions ongoing that we're simply unaware of. Yet there are efforts and laws in place to protect *endangered* and *threatened species*.

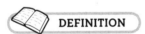 **DEFINITION**

> **Endangered species** are those in danger of becoming extinct due to their low populations. **Threatened species** are likely to become endangered in the foreseeable future.

In the United States, the first calls for species protection began with the near loss of the American bison and the complete disappearance of the passenger pigeon. Before this, the American public was largely unaware of the concept of extinction.

# The Endangered Species Act

The Endangered Species Act was passed in 1973 under Richard Nixon, during the same time period that saw the emergence of the EPA, the Clean Water Act, and the Clean Air Act. To be considered for listing, a species is evaluated based on these criteria:

- The species' habitat has been significantly reduced or threatened.

- The species has been overharvested for any purpose.

- The population of the species is declining.

- There aren't any existing regulations that protect the species.

- Specific natural or man-made factors limit its population.

The law prohibits any action that causes the "taking" of any listed species. This includes hunting, trapping, buying, selling, importing, or exporting.

Included within this idea of "taking" is the destruction of the species' habitat. The Endangered Species Act defines "critical habitat" that's necessary for the recovery of the species. Any damage or destruction of this critical habitat is unlawful.

# The Red List

One shortcoming of the Endangered Species Act is that the process of listing a species can become very politicized. Since the law was signed, the rate at which species were added to the list increased until the presidency of George W. Bush, between 2000 and 2008, when it encountered a sharp decline.

The IUCN Red List is considered the most comprehensive inventory of the conservation status of species of plants, animals, and other organisms. This is only a list, however, and not law.

Internationally, an agreement exists called the Convention on International Trade in Endangered Species or CITIES. Participation in this convention is voluntary, but countries that have signed on have agreed to regulate the taking and trade of any of the species listed.

Each of these laws, measures, and international agreements represents a significant movement forward from the mass of extinctions and near-extinctions of the nineteenth and early twentieth centuries. Yet a great deal of work remains to be done. Deforestation is still rampant in many countries of the world, especially developing ones. Poaching is also a big problem, particularly in Asia, where a lucrative market for animal products persists.

Given time, continued public awareness, and a push for an even greater degree of change, each of our environmental problems can be managed. The question is, how long will it take? The answer lies with you—as a voter, consumer, and member of the public. Governments and world markets ultimately answer to the demands of the public—their constituents and customers. Only when change is demanded will it be given.

## The Least You Need to Know

- Plant and animal extinctions are occurring at an accelerating rate, primarily due to human influences.
- The major causes behind species extinctions are ecosystem degradation, invasive species, and overhunting.
- National parks and preserves set aside land that can be used only for recreational purposes, disallowing any hunting and logging.
- The Endangered Species Act and CITIES regulate any hunting or trade of species in danger of extinction.

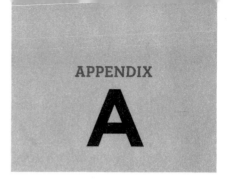

# Glossary

**altitude**   The height above sea level.

**anthropocentrism**   An environmental philosophy that places human needs before those of other animals or ecosystems.

**antinatalist forces**   Factors that decrease the likelihood of an individual having children.

**aquifer**   An isolated source of water that exists below the zone of saturation.

**beef cattle**   Breeds of cows selected for their size and growth rate.

**benthic zone**   The bottom layer of any body of water, where no sunlight is present.

**bias**   The manipulation of data from an experiment on the part of a scientist or a patient.

**bioaccumulation**   The gradual build-up of a toxic substance in the tissues of an organism due to continued exposure through food, water, or air.

**biocentrism**   An environmental philosophy that places the needs of nonhuman species as a priority.

**biodegradable**   Waste or other materials that can be naturally broken down by decomposers in the environment.

**biological oxygen demand (BOD)**   A measurement of how much of the dissolved oxygen in an area of water is being consumed.

**biomagnification**   An increasing concentration of a toxic substance through a food chain, peaking at the top levels.

**biome**   An area of land that has similar soil, climate, and living organisms.

**birth rate**   The number of new individuals born into a population over a given time.

**boom**   A temporary floating barrier that's laid down during an oil spill to prevent surface oil from spreading.

**by-catch**   Fish that are unwanted and unintentionally caught by a trawl net.

**carcinogen**   A mutagenic agent that's known to increase the risk of cancer.

**carrying capacity**   The maximum population size that can be sustained in an ecosystem.

**commensalism**   A symbiotic relationship where one species benefits, while the other is not affected.

**consumer**   An organism that cannot produce its own food directly and must ingest or absorb other living tissue.

**control rods**   Neutron-absorbing material that's inserted between fuel rods in nuclear fission reactors.

**controlled burn**   An oil spill control method that intentionally burns off oil floating on the surface of water.

**dairy cattle**   Breeds of cows specifically selected for the highest milk production possible.

**dead zone**   An area of an aquatic ecosystem with lower-than-normal biodiversity.

**death rate**   The number of individuals in a population that die over a given time.

**desertification**   The gradual conversion of an arid ecosystem into a desertlike state, including low biodiversity and dry soil.

**dieback**   A sudden drop in population that occurs when the carrying capacity of an ecosystem is exceeded.

**dispersant**   A chemical that breaks up oil into smaller droplets, which can disperse much more easily throughout the water column.

**dissolved oxygen (DO)**   A measurement of the amount of oxygen gas in an aquatic ecosystem.

**ecocentrism**   An environmental philosophy that places the well-being of entire ecosystems as a priority.

**emigration**   The number of individuals that move out of an area over a given time.

**Environmental Protection Agency (EPA)**   A U.S. government agency that enforces regulations from the Clean Air Act, Clean Water Act, and other environmental legislation.

**equator**   A line drawn across the Earth directly between the North and South Poles. Designated as 0° latitude.

**erosion**   The movement of soil through forces of wind or water.

**eukaryote**   An organism made of cells that have membrane-bound organelles, such as a nucleus or mitochondria.

**euphotic zone**   The top layer of a body of water, where enough sunlight penetrates to allow photosynthesis to occur.

**evaporation**   When liquid water is heated and becomes water vapor, entering the atmosphere.

**exponential growth**   A pattern of population growth that occurs when the initial growth rate is very slow, but then begins to increase rapidly in an exponential pattern ($2^2$, $2^3$, $2^4$, $2^5$, ...).

**extinction**   The complete disappearance of a species from the Earth.

**fly ash**   The component of ash that actually becomes airborne in the process of incineration.

**free trade**   A label given to brands that meet standards promoting sustainable practices and providing fair compensation.

**freshwater**   A type of aquatic ecosystem where the water isn't salty, such as a river.

**fuel rods**   Stacks of pelletized uranium fuel used in nuclear fission reactors.

**glacial period**   An interval of time within an ice age when overall global temperatures are colder and glaciers increase in size.

**gully erosion**   Small channels carved into soil by running water that gradually enlarge into ditchlike gullies.

**habitat**   The range of physical area or ecosystem that a species resides in.

**habitat fragmentation**   The division of a larger ecosystem into two smaller ones by way of a physical barrier.

**homogenization**   The process of mixing and processing the fat in milk so it's evenly distributed and doesn't separate.

**ice age**   Long periods of glacial movement (millions of years).

**immigration**   The number of new individuals that move into an area over a given time.

**incineration**   The process of burning municipal solid waste to reduce its volume.

**infectious**   A disease that can be spread between individuals.

**infiltration**   When precipitation falls and soaks deep into the soil.

**interglacial period**    An interval of time within an ice age when overall global temperatures are warmer and glaciers decrease in size.

**kinetic energy**    Energy an object possesses due to its motion.

**latitude**    The distance north or south of the equator, measured in degrees.

**LEED certification**    Awarded to buildings that can meet specific standards involving energy efficiency and usage of green materials.

**leeward slope**    The side of a mountain range opposite the prevailing winds that receives proportionately less precipitation.

**marine**    A type of aquatic ecosystem that contains saltwater, such as the ocean.

**moderator solution**    A liquid (usually water) that surrounds the nuclear core, keeping it from overheating.

**MSC certification**    Recognition given to fisheries that agree to limit their annual harvests of fish to more sustainable levels.

**mutagen**    A chemical or type of energy that can damage the DNA of living cells.

**mutualism**    A symbiotic relationship where both species benefit.

**National Oceanic and Atmospheric Administration (NOAA)**    A U.S. government agency that monitors conditions of the oceans and atmosphere, including weather patterns, sea level, and ocean temperatures.

**National Parks Service (NPS)**    A U.S. government agency that manages all national parks and federally designated monuments and historical landmarks.

**natural capital**    All of the resources available from the Earth; everything from fish to timber to clean water.

**natural interest**    A growth in the amount of a given resource on the Earth, such as when a fish population increases.

**niche**    The role an organism occupies within an ecosystem.

**nondegradable**    Waste or other materials that can't be broken down by natural processes.

**nonrenewable resource**    A substance that, once consumed, won't be replenished within a human lifetime.

**North Pole**    The northernmost point of the Earth. Designated with a latitude of 90 degrees N.

**nuclear fission**    The splitting of a Uranium-235 atom, which releases energy.

**overshoot**    When exponential growth carries a population size past the carrying capacity of an ecosystem.

**oxygen sag**   A sudden drop in dissolved oxygen levels along an aquatic ecosystem.

**parabolic solar collection**   The use of mirrors to focus sunlight to generate heat, steam, and eventually electricity.

**parasitism**   A symbiotic relationship where one species benefits at the expense of the other.

**passive solar heating**   Structures designed to capture sunlight for heat without the use of moving parts.

**pasteurization**   A heating process used in milk processing that kills bacteria and maximizes shelf life.

**persistence**   The property of not being easily broken down in the environment once released; most likely to accumulate.

**photovoltaic cell**   Materials that absorb sunlight and directly generate an electric current.

**pole-caught**   A method of commercial fishing in which fish are pulled directly from the ocean by teams of fishermen using poles with hooked lures on a stationary ship.

**potential energy**   Stored energy that exists in certain molecules or as the result of the position of an object.

**precipitation**   Water as it condenses and falls back to the Earth.

**primary pollutant**   A contaminant that's released directly into the air.

**producer**   Any organism that can produce its own molecules of high-potential energy from a source like sunlight or heat.

**prokaryote**   An organism made of cells that lack a nucleus or any other membrane-bound organelles.

**pronatalist forces**   Factors that increase the likelihood of an individual having children.

**pseudoscience**   A claim that appears or presents itself to be objective science, but is not.

**qualitative data**   Non-numerical observations like color change or how strong something tastes.

**quantitative data**   Numerical measurements taken as part of a scientific experiment.

**rainshadow effect**   The impact a mountain range has on precipitation levels in the biomes surrounding it.

**reactor core**   The enclosed part of a nuclear power plant that contains the uranium fuel. The fission chain reaction occurs here.

**renewable resource**   A substance that can be replenished within a human lifetime.

**runoff**    Precipitation that doesn't infiltrate into the soil, but rather moves across the soil and into a body of water or river.

**salinization**    A buildup of salt ions in soil, such as sodium, magnesium, or calcium, usually due to rapid water evaporation

**sanitary landfill**    A method of disposing of municipal waste by burying it in a securely lined pit.

**secondary pollutant**    An air contaminant that forms as a result of a chemical reaction between a primary pollutant and sunlight or water.

**sheet erosion**    An entire layer of topsoil is removed by wind or water.

**skimmer**    An oil spill control tool that physically separates water from oil along the surface of water.

**South Pole**    The northernmost point of the Earth. Designated with a latitude of 90 degrees S.

**symbiosis**    A close relationship between two species.

**teratogen**    A mutagenic agent that is known to increase the risk of birth defects.

**transpiration**    Movement of water out of the leaves of plants and into the air.

**trawling**    A fishing method in which a large mesh net is dragged through the middle or bottom layer of the ocean, catching any species larger than the mesh.

**United States Forest Service (USFS)**    A U.S. government agency that administers publicly owned forests and grasslands.

**waste stream**    The total production of waste from human society, including agricultural, industrial, municipal, and mining sources.

**water table**    The border between the zone of aeration and saturation in soil.

**windward slope**    The side of a mountain range facing the prevailing winds that receives proportionately greater precipitation.

**zone of aeration**    The upper layer of soil, where the empty spaces within are most often filled by air.

**zone of saturation**    The lower layer of soil, where the empty spaces are completely filled by water.

# Resources

## Books

Blackwell, Andrew. *Visit Sunny Chernobyl: And Other Adventures in the World's Most Polluted Places.*
New York: Rodale, 2012.

Davis, Devra Lee. *When Smoke Ran Like Water: Tales of Environmental Deception and the Battle Against Pollution.* New York: Basic Books, 2002.

Diamond, Jared M. *Collapse: How Societies Choose to Fail or Survive.* London: Penguin, 2006.

Gibbs, Lois Marie. *Love Canal: The Story Continues ….* Gabriola Island, B.C.: New Society Publishers, 1998.

Hawley, T. M. *Against the Fires of Hell: The Environmental Disaster of the Gulf War.* New York: Harcourt Brace Jovanovich Publishers, 1992.

Hesser, Leon F. *The Man Who Fed the World: Nobel Peace Prize Laureate Norman Borlaug and His Battle to End World Hunger: An Authorized Biography.* Dallas: Durban House Pub. Co., 2006.

Hvistendahl, Mara. *Unnatural Selection: Choosing Boys over Girls, and the Consequences of a World Full of Men.* New York: PublicAffairs, 2011.

Lapierre, Dominique, and Javier Moro. *Five Past Midnight in Bhopal.* New York: Warner Books, 2002.

Miller, Richard L. *Under the Cloud: The Decades of Nuclear Testing.* New York: Free Press, 1986.

Osif, Bonnie A., Anthony John Baratta, and Thomas W. Conkling. *TMI 25 Years Later: The Three Mile Island Nuclear Power Plant Accident and Its Impact.* University Park, PA: Pennsylvania State University Press, 2004.

Ott, Riki. *Not One Drop: Betrayal and Courage in the Wake of the* Exxon Valdez *Oil Spill.* White River Junction, VT: Chelsea Green Pub., 2008.

Rathje, William L., and Cullen Murphy. *Rubbish!: The Archaeology of Garbage.* New York: HarperCollins Publishers, 1992.

Veron, J. E. N. *A Reef in Time: The Great Barrier Reef from Beginning to End.* Cambridge, MA: Belknap Press of Harvard University Press, 2008.

Williams, Michael. *Deforesting the Earth: From Prehistory to Global Crisis.* Chicago: University of Chicago Press, 2003.

# Documentaries

*An Inconvenient Truth.* Dir. Albert Gore. Perf. Al Gore. Paramount, 2006.

*Climate Refugees.* Dir. Michael Nash. Climate Refugees, distributed by the Video Project, 2010.

*Food, Inc.* Dir. Robert Kenner. Magnolia Home Entertainment, 2009.

*Fresh.* Dir. Ana Sofia Joanes. Ripple Effect Productions, 2009.

*Planet Earth.* Dir. Alastair Fothergill. Discovery Communications, Inc., 2007.

*The Blue Planet.* Dir. Alastair Fothergill. BBC Video, 2002.

# Websites

**BBC News Science and Environment**
bbc.co.uk/news/science_and_environment

**Clean Air Act | US EPA**
epa.gov/air/caa

**Environmental History Timeline**
environmentalhistory.org

**IUCN Red List of Threatened Species**
iucnredlist.org

*New York Times*—Science News—Environment
nytimes.com/pages/science/earth/index.html

**Superfund | US EPA**
epa.gov/superfund

**UNESCO World Heritage Sites**
whc.unesco.org/en/list

# Index

# X-Y-Z